AF179576

MICROBIOLOGY RESEARCH ADVANCES

MOLECULAR BIOLOGY OF BACTERIA

MICROBIOLOGY RESEARCH ADVANCES

Additional books in this series can be found on Nova's website
under the Series tab.

Additional e-books in this series can be found on Nova's website
under the e-book tab.

BACTERIOLOGY RESEARCH DEVELOPMENTS

Additional books in this series can be found on Nova's website
under the Series tab.

Additional e-books in this series can be found on Nova's website
under the e-book tab.

MICROBIOLOGY RESEARCH ADVANCES

MOLECULAR BIOLOGY OF BACTERIA

RAJARSHI KUMAR GAUR
AND
HEMANT K. GAUTAM
EDITORS

New York

Copyright © 2013 by Nova Science Publishers, Inc.

All rights reserved. No part of this book may be reproduced, stored in a retrieval system or transmitted in any form or by any means: electronic, electrostatic, magnetic, tape, mechanical photocopying, recording or otherwise without the written permission of the Publisher.

For permission to use material from this book please contact us:
Telephone 631-231-7269; Fax 631-231-8175
Web Site: http://www.novapublishers.com

NOTICE TO THE READER

The Publisher has taken reasonable care in the preparation of this book, but makes no expressed or implied warranty of any kind and assumes no responsibility for any errors or omissions. No liability is assumed for incidental or consequential damages in connection with or arising out of information contained in this book. The Publisher shall not be liable for any special, consequential, or exemplary damages resulting, in whole or in part, from the readers' use of, or reliance upon, this material. Any parts of this book based on government reports are so indicated and copyright is claimed for those parts to the extent applicable to compilations of such works.

Independent verification should be sought for any data, advice or recommendations contained in this book. In addition, no responsibility is assumed by the publisher for any injury and/or damage to persons or property arising from any methods, products, instructions, ideas or otherwise contained in this publication.

This publication is designed to provide accurate and authoritative information with regard to the subject matter covered herein. It is sold with the clear understanding that the Publisher is not engaged in rendering legal or any other professional services. If legal or any other expert assistance is required, the services of a competent person should be sought. FROM A DECLARATION OF PARTICIPANTS JOINTLY ADOPTED BY A COMMITTEE OF THE AMERICAN BAR ASSOCIATION AND A COMMITTEE OF PUBLISHERS.

Additional color graphics may be available in the e-book version of this book.

Library of Congress Cataloging-in-Publication Data

Molecular biology of bacteria / [edited by] Rajarshi Kumar Gaur (Department of Science, Faculty of Arts, Science and Commerce, Mody Institute of Technology and Science, Rajasthan, India) and Hemant K. Gautam (Institute of Genomics and Integrative Biology, Delhi, India).
pages cm
Includes bibliographical references and index.
ISBN 978-1-62618-189-2 (hardcover)
1. Bacterial genetics. 2. Bacterial genomes. 3. Bacteria. I. Gaur, Rajarshi Kumar, editor of compilation. II. Gautam, Hemant K.., editor of compilation.
QH434.M638 2013
572.8'6293--dc23 2013005769

Published by Nova Science Publishers, Inc. † New York

Contents

Preface

Bacteria are not just important as laboratory tools to understand higher organisms; they are important and interesting in their own right. For instances, they play an essential role in the ecology of earth. They are the only organisms that can fix atmospheric nitrogen, that is convert N_2 to ammonia, which can be used to make nitrogen-containing cellular constituents such as proteins and nucleic acids. Moreover bacteria produce most of the naturally occurring so-called greenhouse gases, such as methane and carbon dioxide, which are in turn used by other types of bacteria.

Bacteria are also worth studying because of their role in disease. They cause many human, plant and animal diseases and new disease are continuously appearing. Knowledge gained from the molecular genetics of bacteria will help in the development of new ways to treat or otherwise control old diseases, as well as new ones.

Molecular Biology has proved to be one of the more fruitful technological approaches sciences, being both very powerful and able to generate valuable intellectual property. This book aims to present examples in the application molecular biology and genetic engineering in bacteriology. This book shows diverse roles of bacteria in the ecosystems and it gives significant contributions from biotechnology approaches. The challenges of these aspects and methodologies inspired us to edit the book entitled *Molecular Biology of Bacteria.*

Rajarshi Kumar Gaur
and
Hemant K. Gautam
Editors

About the Editors

Dr. R.K. Gaur is presently working as Head and Assistant Professor, Department of Science, Mody Institute of Technology and Science (Deemed University), Lakshmangarh, Sikar, Rajasthan. He did his Ph.D on molecular characterization of sugarcane viruses of India. He partially characterized three sugarcane virus viz., sugarcane mosaic virus, sugarcane streak mosaic virus and sugarcane yellow luteovirus. He received MASHAV fellowship in 2004 of Israel government for his post-doctoral studies and joined The Volcani Centre, Israel and then shifted to Ben Gurion University, Negev, Israel. In 2007 he received the Visiting Scientist Fellowship from Swedish Institute Fellowship, Sweden for one year to work in the The Umeå University, Umeå, Sweden. He is also a recipient of ICGEB, Italy Post-Doctoral fellowship in 2008. He worked on development of marker-free transgenic plant against cucumber viruses. He has made significant contributions on plant microbes and published 56 national/international papers and presented near about 45 papers in the national and international conferences. He has also visited Thailand, New Zealand, London and Italy for the sake of attending the conference/workshop. He is also a member of national and international societies.

Dr. Hemant K. Gautam obtained his Masters and Ph.D. degree in microbiology from the Indian Agricultural Research Institute, New Delhi. He did post doctorate from France and Israel. He is currently working as a Principal Scientist at Institute of Genomics and Integrative Biology (CSIR), Delhi. He is associated with number of Universities and members of various scientific and academic organizations. He is a recipient of Bharat Excellence award, Biotechnology award, IFP France & Israel Govt. fellowship and UNESCO fellow. He published more than 60 research papers and over 75 new sequences have been submitted to NCBI database. He has a distinguish background in the field of microbial biotechnology and natural products.

List of Contributors

Arinthip Thamchaipenet	Actinobacterial Research Unit, Department of Genetics, Faculty of Science, Kasetsart University, Chatuchak, Bangkok 10900, Thailand Center for Advanced Studies in Tropical Natural Resources, National Research University-Kasetsart University (CASTNAR, NRU-KU), Chatuchak, Bangkok 10900, Thailand
Chakrit Bunyoo	Actinobacterial Research Unit, Department of Genetics, Faculty of Science, Kasetsart University, Chatuchak, Bangkok 10900, Thailand
Ratchaniwan Jaemsaeng	Actinobacterial Research Unit, Department of Genetics, Faculty of Science, Kasetsart University, Chatuchak, Bangkok 10900, Thailand Center for Advanced Studies in Tropical Natural Resources, National Research University-Kasetsart University (CASTNAR, NRU-KU), Chatuchak, Bangkok 10900, Thailand
Karan Lomaneeratana	Actinobacterial Research Unit, Department of Genetics, Faculty of Science, Kasetsart University, Chatuchak, Bangkok 10900, Thailand
Siriwan Rungin	Actinobacterial Research Unit, Department of Genetics, Faculty of Science, Kasetsart University, Chatuchak, Bangkok 10900, Thailand
Somkiat Phornphisutthimas	Department of Biology, and Research Unit on Science, Technology and Environment for Learning, Faculty of Science, Srinakharinwirot University, Sukhumvit 23, Bangkok 10110, Thailand

Pankaj Kumar Jain	Biotechnology Laboratory, Department of Science, MITS University, Lakshmangarh, Sikar-332311, Rajasthan, India
Vivek Bajpai	Department of Microbiology, ITM University, Gwalior, MP, India
Madan Lowry	Center for advance studies, Department of Zoology, University of Rajasthan, Jaipur- 302004, Rajasthan, India
Rajal Debnath	Biotechnology Division, CSIR- North East Institute of Science & Technology,Jorhat-785006, Assam, India
Rupak K. Sarma	Biotechnology Division, CSIR- North East Institute of Science & Technology, Jorhat-785006, Assam, India
Ratul Saikia	Biotechnology Division, CSIR- North East Institute of Science & Technology, Jorhat-785006, Assam, India
Archana Yadav	Biotechnology Division, CSIR- North East Institute of Science & Technology, Jorhat-785006, Assam, India
Tarun C Bora	Biotechnology Division, CSIR- North East Institute of Science & Technology, Jorhat-785006, Assam, India
Kashyap Kumar Dubey	Microbial Biotechnology Laboratory, University Institute of Engineering and Technology, M.D. University Rohtak, Rohtak-124001 Haryana, India
Dhirendra Kumar	Microbial Biotechnology Laboratory, University Institute of Engineering and Technology, M.D. University Rohtak, Rohtak-124001 Haryana, India
Anamika Jha	Ashok and Rita Patel Institute of integrated Study and Research in Biotechnology and Allied Sciences, New Vallabh Vidyanagar,Anand- 388121 (Gujarat), India
Hemant K. Gautam	Institute of Genomics and Integrative Biology Mall Road, University Campus, Delhi-110 007 India
Nisha Daxini	Ashok and Rita Patel Institute of integrated Study and Research in Biotechnology and Allied Sciences, New Vallabh Vidyanagar, Anand- 388121 (Gujarat), India
Sakshi Issar	Department of Science, faculty of Arts, Science and Commerce, Mody Institute of Technology and Science- Lakshmangarh, Sikar-332311, Rajasthan, India

D.K.Choudhary	Department of Science, faculty of Arts, Science and Commerce, Mody Institute of Technology and Science-Lakshmangarh, Sikar-332311, Rajasthan, India
R.K.Gaur	Department of Science, faculty of Arts, Science and Commerce, Mody Institute of Technology and Science-Lakshmangarh, Sikar-332311, Rajasthan, India
Komal Saxena	Institute of Biomedical Education & Research, Mangalayatan University, Aligarh – Uttar Pradesh, India
Girijesh Kumar Patel	Department of Biotechnology, Graphic Era University, 566/6, Bell Road, Clement, Town, Dehradun, Uttarakhand, India
G.K. Aseri	Amity Institute of Biotechnology, Amity University Rajasthan, Jaipur, India
Vikash Babu	Department of Biotechnology, Graphic Era University, 566/6, Bell Road, Clement, Town, Dehradun, Uttarakhand, India
Chand Ram	Bioremediation Laboratory, Dairy Microbiology division, National Dairy Research Institute, Karnal-132001 (Haryana), India
Vijay Kumar	Bioremediation Laboratory, Dairy Microbiology division, National Dairy Research Institute, Karnal-132001 (Haryana), India

In: Molecular Biology of Bacteria
Editors: R. Kumar Gaur and H. K. Gautam
ISBN: 978-1-62618-189-2
© 2013 Nova Science Publishers, Inc.

Chapter 1

Intergeneric Conjugation: A Practical Method for Genetic Manipulation in Actinomycetes

Arinthip Thamchaipenet[1,2*], Chakrit Bunyoo[1], Ratchaniwan Jaemsaeng[1,2], Karan Lohmaneeratana[1], Siriwan Rungin[1] and Somkiat Phornphisutthimas[3]

[1]Actinobacteria Research Unit, Department of Genetics, Faculty of Science, Kasetsart University, Chatuchak, Bangkok, Thailand
[2]Center for Advanced Studies in Tropical Natural Resources, National Research University-Kasetsart University (CASTNAR, NRU-KU), Chatuchak, Bangkok, Thailand
[3]Department of Biology, and Research Unit on Science, Technology and Environment for Learning, Faculty of Science, Srinakharinwirot University, Bangkok, Thailand

Abstract

The system for intergeneric conjugation has been developed for many actinomycete species including *Streptomyces* and rare actinomycetes. *E. coli* donor strains generally used are the methylation proficient strain, S17-1 harbouring *tra* genes on its chromosome and the methylation-deficient strain, ET12567 containing *tra* genes in a self-transmissible plasmid pUB307 or a non-transmissible plasmid pUZ8002. Integrative plasmids used are normally non-replicative plasmids carrying the *oriT* for the propose of gene disruption (e.g. pSET151, pIJ8671) or with additional ϕC31 *attP-int* system in the case of gene expression (e.g. pSET152, pIJ8600). Actinomycete recipients used for mating with *E. coli* could be in spore suspension with or without pre-germination process, or mycelia in liquid culture. Several agar media including MS and ISP 4 were generally used for

* Correspondence: Arinthip Thamchaipenet, Actinobacteria Research Unit, Department of Genetics, Faculty of Science, Kasetsart University, Chatuchak, Bangkok 10900, Thailand. Tel.: +66 2 562 5444 ext. 4208; fax.: +66 2 579 5528; e-mail: arinthip.t@ku.ac.th.

streptomycete strains while other more specific media were proposed for rare actinomycetes. Concentration of $MgCl_2$, ratio of donor and recipient and other parameters are significantly important to increase conjugation efficiency. The intergeneric conjugation approach has become a powerful method to genetically manipulate actinomycete species with high efficiency. It turns out to be a common tool to characterise functions of genes/gene clusters in these pharmaceutically important bacteria.

Introduction

Actinomycetes are high G+C content (74 mol%) filamentous Gram-positive bacteria found abundantly in soil, sand and sediments. They have a complex life cycle with an extended substrate mycelia on which aerial hyphae is mostly formed with exospores. Genus *Streptomyces* in particular is a key producer of valuable secondary metabolites including anti-bacterials, anti-fungals, anti-cancers, immunosuppressants, herbicides and insecticides used nowadays in medicine, pharmaceutics and agriculture [1, 2]. Actinomycetes, therefore, are the most attractive microorganisms to study genes involved in biosynthesis of such important secondary metabolites. Recently, genome analysis of *Streptomyces* chromosomes revealed that they carry vast numbers of uncharacterised genes and biosynthetic gene clusters for undiscovered secondary metabolites [3]. Therefore, the establishment of efficient genetic manipulation systems to introduce homologous/heterologous DNAs into actinomycete species has become a critical requirement for achieving functional analysis of the genes/gene clusters of interest.

Genetic manipulation in actinomycetes was developed intensively for the genus *Streptomyces*. Recombinant plasmids construction for gene expression and disruption/replacement can be generally transferred into streptomycete hosts using polyethylene glycol (PEG) mediated protoplast [4]. The procedure is based on the protocol defined for *S. coelicolor* and *S. lividans* [5]. However, this method is time-consuming procedure as it requires high productivity of protoplast formation and regeneration to achieve good transformation efficiency. With a number of different streptomycete species, the method has to be optimised e.g. mycelial age, concentration of the lysozyme to give good protoplast formation and conditions for protoplast regeneration. PEG mediated protoplast transformation does not easily applicable to every species and frequently results in low efficiency particularly with rare actinomycete strains. In addition, due to the strong host restriction modification system of *Streptomyces* [6], it is essential to find a proficient method to transfer DNA with high efficiency. Although high efficient transformation by electroporation was proposed to avoid the problematic protoplast formation and regeneration [7, 8], it requires luxurious equipment and has limited application not to every species of actinomycetes.

Nowadays, the method generally employed for genetic manipulation in *Streptomyces* and other genera is plasmid-mediated intergeneric conjugation with *E. coli* as a donor [9]. This method is proved to be simple and less time consuming and likely to be applied to various actinomycete species. However, the efficiency of this system varies significantly from one species to another, and it is necessary to improve experimental procedures for a new species.

Bacterial Conjugation

Conjugation is a common phenomenon in both Gram-negative and Gram-positive bacteria. It requires close physical contact between donor and recipient cells and is usually mediated by plasmid-encoded proteins which provide transfer (*tra*) functions [10, 11, 12].

The conjugation by self-transmissible plasmid is related in a mechanism sense to that of F plasmid in *E. coli*. Donor creates F pilus to make cell-to-cell contact to recipient. The mechanism is initiated by creating a single-stranded nick at the origin of transfer (*oriT*). The 5'-end of the nicked DNA is transported linearly through the F pilus into the recipient cell where it is replicated; while, the unnicked DNA is replicated in the donor cells [13].

Naturally, some self-transmissible plasmids could transfer among the same bacterial species but not to other different species. However, some plasmids were able both to transfer and to replicate in effectively different species. Some conjugation phenomena with non-pilus mechanisms were also found in Gram-positive bacteria including *Streptomyces* [10].

Conjugation within *Streptomyces* Species

The conjugation process with non-pilus mechanism naturally appears in *Streptomyces*. It is a very distinct process from other bacteria in its molecular mechanism and its visible phenotypic appearance. It occurs through intermycelial transfer from the hyphal tips of a plasmid carrying donor to recipient [14, 15]. It seems that such single-stranded DNA transfer that appears in all conjugation systems in unicellular bacteria does not involve in *Streptomyces* conjugation.

Conjugative plasmid was transferred as a double-stranded DNA molecule [15, 16] by *tra* gene products, e.g. *traB* for pSN22 [17] and *traSA* for pSAM2 [16] during mating [18]. Then the plasmid is intramycelial spread through a septum to the next cell of the recipient mycelium. At this point, the growth is retarded in order to obtain sufficient copy number of the plasmids for the next spreading to another compartment of the mycelium [14, 19]. This delay causes a pock formation, a visible retardation zone of plasmid-acquiring recipient cells grown in a confluent lawn of plasmid-free recipients [20, 21].

Intergeneric Conjugation between *E. coli* and *Streptomyces*

Conjugal transfer from *E. coli* to Gram-positive bacteria with F plasmid-like system was demonstrated by Trieu-Cuot et al. [22]. In 1989, intergeneric conjugation from *E. coli* to *Streptomyces* was firstly described [9] and later has been developed and applied to several genera of *Actinomycetales*. This method has proved to be easy, convenient and less time-consuming compared to the original PEG-mediated protoplast transformation method.

The system of intergeneric conjugation between these two taxonomically different microorganisms is absolutely required an *oriT* and *tra* genes for transfer functions in the donor *E. coli* host. It is hypothetically explained that a plasmid was transferred from *E. coli* to *Streptomyces* by single-stranded DNA forms generated by *oriT*-endonuclease function of donor [10]. This transfer method then could overcome problems of the restriction barrier to foreign DNA in *Streptomyces* hosts by the transfer of single-stranded DNA [23].

E. coli Donor and Mobilisable Plasmids

E. coli donor strains generally used for intergeneric conjugation are the methylation proficient strain, S17-1 harbouring RP4 *tra* genes on its chromosome [9], the methylation-deficient strain, ET12567 (*dam-13::Tn9 dcm-6 hsdM*) [24] containing *tra* genes in a self-transmissible plasmid pUB307 [23] and ET12567 carrying a non-transmissible plasmid pUZ8002 [25]. pUZ8002, a helper plasmid RK2 derivative, is not efficiently transferred itself because of a mutation in its own *oriT* but can supply transfer functions to mobilise *oriT*-carrying non-conjugative replicon plasmids [26]. By using methylation deficient donor such *E. coli* ET12567, the restriction barriers of foreign DNA could be reduced and efficiency of DNA introduction could be highly increased in *Streptomyces* host [6, 23].

The plasmids used for intergeneric conjugal transfer are mobilisable plasmids that required *tra* genes function of F plasmid located on either host chromosome (e.g. *E. coli* S17-1) or helper plasmid (e.g. pUZ8002). There is no evidence yet whether a pilus mechanism is required for mating pair formation between *E. coli* and *Streptomyces*.

Non-replicative plasmids are used on propose of integration of genetic materials into *Streptomyces* chromosomes. These plasmids carry the origin of replication for *E. coli* and the *oriT* region of the broad host range conjugative plasmid RK2 [9, 10]. If the plasmids required for gene expression propose, integrase gene (*int*) and attachment site (*attP*) of *Streptomyces* phage ϕC31 were inserted in order to drive site-specific recombination at *attB* site of *Streptomyces* chromosome [27, 28]. These plasmids, for example, are pIJ8600 (*ori*pUC, *oriT*, ϕC31 *attP-int*, Apr^r, Thio^r, *tipA*p) [29], pSET152 (*ori*pUC, *oriT*, ϕC31 *attP-int*, Apr^r) [28], pIJ6902 (*ori*pUC, *oriT*, ϕC31 *attP-int*, Apr^r, Thio^r, *tipA*p) [30] and pIJ10257 (*ori*pUC, *oriT*, ϕBT1 *attB-int*, Hyg^r, *ermE*p*) [31]. If the plasmids are to use for integration by homologous recombination, no *int* and *attP* of ϕC31 was required e.g. pIJ8671 (*ori*pUC, *oriT*, Apr^r, Thio^r, *tipA*p) [29] and pSET151 (*ori*pUC, *oriT*, Thio^r) [32].

The non-replicative recombinant plasmid is forced to integrate at the target homologous gene to achieve disruption. Some shuttle plasmids were designed for replication purpose in *Streptomyces* then the origin of replication for both *E. coli* and *Streptomyces* was inserted such as pJN100 (*ori*pUC, *ori*pIJ101, *oriT*, Apr^r, *snpA*p) [33].

Pre-Treatments for Streptomycete-Actinomycete Recipients

Pre-germination of spores of *Streptomyces* recipients by heat treatment is critical for intergeneric conjugation procedure [5, 23]. The temperature and time used for spore pre-germination treatment effect on the efficiency of conjugation in individual species. In general, pre-germination of *Streptomyces* spores activated by heat treatment at 50°C for 10 min before mating with an *E. coli* donor was recommended [5, 9, 23]. The spore pre-germination of *S. lividans* at this condition increased the efficiency of conjugation by 5-10 times [9]. It was practically used for spores of *S. albus*, *S. anibioticus*, *S. aureofaciens*, *S. bambergiensis*, *S.*

fradiae, *S. gresius*, *S. hygroscopicus*, *S. lavendulae*, *S. lividans*, *S. purpureus*, *S. rimosus* ATCC 23955, *S. venezuelae* ATCC 21405, *S. viridochromogenes* [34], *S. avermitilis* [35], *S. clavuligerus* [36], *S. coelicolor* [23], *S. diastatochromogenes* [37], *S. ipomoeae* [38], *S. nodusus* [39], *S. pristinaespiralis* [40], *S. sioyaensis* [41] and a marine *Streptomyces* sp. M095 [42] (Table 1). This condition could, furthermore, use for spore pre-germination of some rare actinomycetes such as *Amycolatopsis japonicum* [37] and *Saccharopolyspora erythraea* [35] (Table 2). However, heat treatment at 50 °C for 10 min was not an optimal condition for pre-germination of spores for every species. To vary the conditions for each strain to maintain good spore viability for high conjugation efficiency, is highly recommended. For instance, requirement of 10 min heat treatment of spores of *S. lavendulae* FRI-5 [43], a rimocidin-producing *S. rimosus* R7 (ATCC 10970) as well as an oxytetracycline-producing *S. rimosus* M4018 [44] was at 40°C, while *S. natalensis* was at 45°C [45].

Some strains required longer pre-germination time such as 50 °C for 25 min for *S. conglobatus*, *S. routienii* [33], *S. nodosus* [46] and *Saccharopolyspora spinosa* [33]. The medermycin-producing *Streptomyces* sp. AM-7161 even required very long 1.5–3.0 h incubation time at 50 °C for optimal spore treatment [47]. In the case of heat-sensitive strains, e.g. *S. virginiae* [34], *Actinoplanes teichomyceticus* [48] and *Kitasatospora setae* [49], spores were used at room temperature without heat treatment. In our current study, most of *Streptomyces* strains achieved a high conjugation efficiency (10^{-2}–10^{-5} exconjugants per recipients) when spores were heat treated at 40°C for 10 min. e.g. *S. rimosus* R7 and M4018 [44], chloramphenicol-producing *S. venezuelae* ATCC 10712 and some endophytic streptomycetes including siderophore-producing *Streptomyces* sp. GMKU 169 and *Streptomyces* sp. GMKU 3100 [50] (Table 1).

However, some strains could give reasonably high conjugation frequency without pre-germination of spores. For example, spores of *S. rimosus* R7 [44], *S. venezuelae* ATCC 10712 and an ACC deaminase-producing endophytic *Streptomyces* sp. GMKU 336 could be used immediately at room temperature without heat treatment (Table 1). There were also some other reports that no heat treatment of spores was required for *S. ambofaciens* [51], *S. albulus* [52] and *S. toyocaensis* [53]. This indicated that the heat treatment step was not prerequisite for every streptomycete species. In contrast, few exconjugants were obtained when spores of *S. pristinaespiralis* were not subjected to the heat treatment [40].

In case of poorly or non-sporulated actinomycete strains, vegetative mycelia could be considered as alternative recipients instead of spores. It is noteworthy that mycelia of such strains must appear in fine particle or fragmented not forming any pellet or clump in liquid culture otherwise mechanical homogenisation or sonication should be applied. Exponential growth (24-h) of mycelia in tryptic soy broth (TSB) of *S. rimosus* R7, M4018 [44], *S. venezuelae* ATCC 10712, endophytic *Streptomyces* sp. GMKU 336 and *Streptomyces* sp. GMKU 3100 [50] appeared naturally dispersed and moderately fragmented under the light microscope. Thus, no mechanical application was needed before mating with *E. coli* as suggested for mycelia of poorly sporulating *S. peucetius* [54].

Table 1. Summary of intergeneric conjugation parameters between *E. coli* and streptomycete strains

Streptomycete strains	Properties	Spore treatment/ mycelium	*E. coli**/plasmid	Agar medium +10 mM $MgCl_2$	Conjugation efficiency	References
S. albus G	*Sal*GI	50°C, 10 min	S17-1/pTO1	MS	5×10^{-4}	[33]
S. ambofaciens ATCC 15154	Spiramycin	No heat (RT)	pSET152	AS-1	1.4×10^{-2}	[51]
S. albulus IFO14147	ε-Poly-L-lysine	No heat (RT)	S17-1/pLAE003	ATCC no. 5 +40 mM $MgCl_2$	4×10^{-7}	[52]
S. antibioticus ATCC 23879	Oleandomycin	50°C, 10 min	S17-1/pTO1	MS	3×10^{-5}	[33]
S. aureofaciens VKPM AC 755	Chlortetracycline				5×10^{-4}	
S. avermitilis ATCC 31272	Avermectin	50°C, 10 min	S17-1	YSP	NA	[34]
S. bambergiensis ATCC 13879	Moenomycin	50°C, 10 min	S17-1/pTO1	MS	1×10^{-5}	[33]
S. clavuligerus ATCC 27064	Clavulanic acid	50°C, 10 min	pALF297	MS	1×10^{-4}	[35]
S. coelicolor A3(2)	Actinorhodin	50°C, 10 min	ET12567 (pUB307)/ pDJ70	MS	1.1×10^{-1}	[23]
S. couglobatus	-	50°C, 25 min	pJN100	MS	1×10^{-6}	[45]
S. diastatochromogenes Tu6028	Polyketomycin	50°C, 10 min	GM2163/SET152	MS	1.6×10^{-4}	[36]
S. fradiae B-45 VKPM AC 542	Tylosin	50°C, 10 min	S17-1/pTO1	MS	1×10^{-4}	[33]
S. griseus Kr.15 VKPM AC 115	Grisin				2×10^{-4}	
S. hygroscopicus ATCC 21705	Bialaphos Hygromycin				1×10^{-3}	
S. hygroscopicus ATCC 10976					1×10^{-3}	
S. ipomoeae 91-03	Sweet potato pathogen	50°C, 10 min	S17-1/pSET152	ISP 4 +20 mM $MgCl_2$	4.1×10^{-5}	[37]
S. lavendulae ATCC 19777	Phytobacteriomycin	50°C, 10 min	S17-1/pTO1	MS	1×10^{-4}	[33]
S. lividans TK64	-				1×10^{-3}	
S. lavandulae FRI-5	γ-Butylolactone IM-2	40°C, 10 min	pSET152	ISP 2	1.6×10^{-5}	[42]
S. lincolnensis ATCC 25466	Lincomycin	3-d mycelium	pUWL201apr	MS +20 mM $MgCl_2$	1.1×10^{-4}	[55]
S. natalensis ATCC 27448	Pimaricin	45°C, 10 min	pHZ1358	MS	8.55×10^{-5}	[44]
S. nodosus ATCC 14899	Amphotericin	50°C, 25 min	pSOK101	MS	5×10^{-5}	[38]
S. peucetius ATCC 29050	Daunorubicin Doxorubicin	Heat sensitive 18-h mycelium	pSET152	MS	1.5×10^{-4}	[54]

**E. coli* ET12567 (pUZ8002) unless indicated, NA = not available.

Streptomycete strains	Properties	Spore treatment/ mycelium	*E. coli**/plasmid	Agar medium $+10$ mM $MgCl_2$	Conjugation efficiency	References
S. pristinaespiralis ATCC 25486	Pristinamycin	50°C, 10 min	pSET152	MS $+30$ mM $MgCl_2$	1.36×10^{-3}	[39]
S. purpureus ATCC 21405	-	50°C, 10 min	S17-1/pTO1	MS	1×10^{-4}	[33]
S. rimosus ATCC 23955	Oxytetracycline	50°C, 10 min	S17-1/pTO1	MS	1×10^{-4}	[33]
S. rimosus R7 (ATCC10970)	Rimocidin	40°C, 10 min	pIJ8600	TSA (Oxoid)	4.21×10^{-3}	[43]
S. rimosus M4018	Oxytetracycline				5.48×10^{-2}	
S. routientii	-	50°C, 25 min	pJN100	MS	2×10^{-6}	[45]
S. sioyaensis NRRL-B5408	Siomycin	50°C, 10 min	ET12567 (pUB307)/pSOK101	ISP 3	1.1×10^{-5}	[40]
S. toyocaensis A80937	Glycopeptide A47934	No heat (RT)	S17-1/pOJ436	R2	2.2×10^{-4}	[53]
S. venezuelae ATCC 10595	Chloramphenicol	50°C, 10 min	S17-1/pTO1	MS	2×10^{-4}	[33]
S. venezuelae ATCC 10712	Chloramphenicol	40°C, 10 min	pSET152	TSA (Oxoid), MS	6.9×10^{-4}	This work
S. virginiae ATCC 13161	Virginiamycin	Heat sensitive (RT)	S17-1/pTO1	MS	2×10^{-3}	[33]
S. viridochromogenes Tu 494	Bialaphos	50°C, 10 min	S17-1/pTO1	MS	3×10–4	[33]
Streptomyces sp. AM-7161	Medermycin	50°C, 1.5-3 h	pSET152	G+Y/GYM	NA	[47]
Streptomyces sp. M095	Holomycin	50°C, 10 min	pIJ8600	MS	1.9×10^{-4}	[41]
Streptomyces sp. GMKU 169	Endophyte/Siderophore	40°C, 10 min	pIJ8600	MS	3.85×10^{-5}	This work
Streptomyces sp. GMKU 336	Endophyte/ACC deaminase	RT	pIJ8671	MS	3.61×10^{-5}	This work
Streptomyces sp. GMKU 3100	Endophyte/Siderophore	40°C, 10 min	pIJ8600	MS	4.5×10^{-4}	[50]
Streptomyces sp. GMKU 944	Endophyte	24-h mycelium	pIJ10257	ISP 4	2.18×10^{-3}	This work

**E. coli* ET12567 (pUZ8002) unless indicated, NA = not available.

Table 2. Summary of intergeneric conjugation parameters between *E. coli* and rare actinomycetes

Actinomycete strains	Properties	Spore treatment/ mycelium	*E. coli**/plasmid	Agar medium +10 mM $MgCl_2$	Conjugation efficiency	References
Actinoplanes teichomyceticus NBCR 13999	Teicoplanin	Heat sensitive (RT)	pSET152	ISP 2 +40 mM $MgCl_2$	4.9×10^{-3}	[48]
Amycolatopsis japonicum DSM 44213	Biodegradable EDTA isomer EDDS	50°C, 10 min	GM2163/pSET152	MS	$<1\times10^{-7}$	[36]
Kitasatospora setae IFO 14216	Bafilomycin B1	No heat (RT)	pSET152	ISP 2 +30 mM $MgCl_2$	1.7×10^{-6}	[49]
Microbispora corallina NRRL 30420	Microbisporicin	48-h mycelium	pSET152	V0.1	NA	[61]
Micromonospora aurantiaca W2b	Polyisoprene rubber degrading	3-d mycelium	pGM446	MS	3.6×10^{-3}	[58]
Micromonospora griseorubida A11725	Mycinamicin II	5-d mycelium	pGM446	MR0.1S	NA	[59]
Micromonospora rosaria IFO13697	Rosamicin	5-d mycelium	pSET152	R2YE	NA	[60]
Micromonospora sp. 40027	Fortimicin A	36-h mycelium	pSET152	Bennet	NA	[57]
Micromonospora sp. GMKU 358	Endophyte/ PLA degradation	36-h mycelium	pIJ8600	V0.1	1.04×10^{-3}	This work
Nonomuraea sp. ATCC 39727	Teicoplanin-like glycopeptide A40926	24-h mycelium	pSET152	V0.1	0.8×10^{-3}	[62]
Saccharopolyspora erythraea ATCC 11635	Erythromycin	50°C, 10 min	S17-1	YSP	NA	[34]
Saccharopolyspora spinosa	Spinocin A/D	50°C, 25 min	pJN100	ABB13	5×10^{-7}	[45]

**E. coli* ET12567 (pUZ8002) unless indicated, NA = not available.

All above strains were able to use either spores or mycelia for intergeneric conjugation and gave equally high efficiency. It resembled *S. fradiae* that using mycelium as a recipient gave no different efficiency from its recipient spores [28]. However, the conjugation of recipient mycelia of *S. toyocaensis* gave lower efficiency than those of recipient spores [53]. It was observed that poorly sporulating endophytic *Streptomyces* sp. GMKU 944 could achieve high conjugation efficiency by using mycelium (Table 1). Similar to a non-spore forming *S. rimosus* R6-593, a mutant of oxytetracycline-producing *S. rimosus* R6, vegetative mycelium was used successfully as recipient instead of spores [44]. On the contrary, an attempt to use mycelium of *S. natalensis* for recipient was unsuccessful [45]. In general, young mycelia (18–24 h) were preferred for intergeneric conjugation [44, 54] but *S. lincolnensis* needs 3-day growth before mating [55].

Utilising mycelia as recipients has been increasingly reported particularly from rare actinomycetes that produced low amounts of spores or none including *Actinomadura*, *Arthrobacter*, *Micromonospora*, *Nocardia*, *Nonomuraea* and *Rhodococcus* [34, 56]. Endophytic *Micromonospora* sp. GMKU 358 carrying poly L-lactic acid (PLA) degradation and fortimicin producing *Micromonospora* sp. 40027 [57] were successfully manipulated by using 36-h mycelial growth as recipient (Table 2). Some species of *Micromonospora* required a longer period of growth of mycelia up to 3-5 days [58, 59, 60] (Table 2). However, optimum condition for *Microbispora corallina* was 48-h mycelial growth [61] while *Nonomuraea* sp. ATCC 39727 required 24-h mycelium [62]. Since mycelia are relatively simple to prepare comparison of spores, it is convincingly advised to use mycelia instead of spores for routinely conjugation experiments for both streptomycetes and non-streptomycetes.

Conjugation Media

One of the important factors affecting the efficiency of intergeneric conjugation is the type of medium used for conjugation. It was demonstrated that conjugation of *E. coli* occurred at a higher frequency on solid than liquid medium [63]. Liquid medium was, however, attempted as a conjugation medium, but it was not successful for intergeneric conjugation between *E. coli* and *Streptomyces* [9]. There were several solid media earlier proposed for intergeneric conjugation such as AS1, LA, R2, SR6, SR12, TSA and YSP [9, 28, 34, 35, 51, 53]. AS1 and R2 agar media were successfully used for *S. ambofaciens* [51], *S. fradiae* [28] and *S. toyocaensis* [53]. However, some of these agar media were inappropriate for conjugation since *E. coli* would outgrow covered the whole plate after overnight incubation. After overlaying with nalidixic acid to get rid of *E. coli*, scraping a lawn of *E. coli* was rather tedious. MS agar medium [64] supplemented with 10 mM $MgCl_2$ was generally recommended as conjugative medium used effectively in many *Streptomyces* species (Table 1) including some non-streptomycetes (Table 2). However, low number or lack of exconjugants sometimes occurred in some streptomycete strains using MS agar as in the cases of *S. ipomoeae* [38] and *S. sioyaensis* [41] where ISP 4 and ISP 3 [65] were respectively replaced.

Although the MS medium supplemented with 10 mM $MgCl_2$ was efficiently used for conjugation experiment in *S. rimosus*, a number of false exconjugants was occasionally obtained due to the soybean component in MS agar that absorbs selective antibiotics and, thus lowers the actual concentration [44]. Therefore, TSA (Oxoid) containing 10 mM $MgCl_2$ was selected and gave 10 fold higher efficiency of conjugation than MS and ISP 4 [44]. Furthermore, TSA (Oxoid) could give equivalent conjugation efficiency to MS agar when used for *S. venezuelae* ATCC 10712 (Table 1). Recent investigation on conjugation media revealed other suitable media for some certain species such as ATCC no. 5 and GYM for *S. albulus* [52] and a medermycin-producing *Streptomyces* sp. AM-7161 [47], respectively.

For rare actinomycetes, finding an appropriate conjugation medium for each species is significantly important. Bennet agar was used for *Micromonospora* sp. 40027 [57], while MR0.1S and R2YE were used for *M. griseorubida* [59] and *M. rosaria* [60], respectively. Conjugative medium giving high efficiency for *M. aurantiaca* was MS (10^{-3}) but low efficiency was observed when used TS and R5 media [58]. However, endophytic *Micromonospora* sp. GMKU 358 revealed very few exconjugants on MS medium but gave high efficiency on V0.1 medium (Table 2) which was used successfully for *Microbispora corallina* [61] and *Nonomuraea* sp. ATCC 39727 [62]. ISP 2 with different concentration of $MgCl_2$ was employed for *Actinoplanes teichomyceticus* and *Kitasatospora setae* [48, 49]. Some species required specific medium such as ABB13 for *Saccharopolyspora spinosa* [33].

Conjugation media are routinely supplemented with 10 mM $MgCl_2$. Although no clear evidence regarding the function of $MgCl_2$, optimal concentration of magnesium is essential for intergeneric conjugation of some certain species [49]. If low conjugation efficiency (10^{-8}–10^{-9}) was obtained using standard protocol [5], concentration of $MgCl_2$ should be optimised to increase the efficiency. It is noteworthy that high concentrations of $MgCl_2$ could severely inhibit growth and spore formation [49]. *S. ipomoeae* required 20 mM $MgCl_2$ in ISP 4 to reach good efficiency [38], while *S. lincolnensis* needed 20 mM $MgCl_2$ in MS [55]. 30 mM $MgCl_2$ was practically used for *S. pristinaespiralis* and *Kitasatospora setae* and could increase 10 times efficiency [40, 49]. *S. albulus* and *Actinoplanes teichomyceticus* required concentration of $MgCl_2$ as high as 40 mM in their conjugal media to achieve desired efficiency [48, 52].

Intergeneric Conjugation Procedure and Related Parameters

Intergeneric conjugation performs by mating an appropriate ratio of an exponential *E. coli* donor (OD_{600} = 0.4–0.6) and recipient spores/mycelia of actinomycetes and selecting the exconjugants on the agar medium with suitable antibiotics [5]. The conjugation efficiency was calculated as the number of exconjugants per recipients. The number of exconjugants per recipients about 10^{-2}–10^{-4} is considered as high efficiency, while 10^{-7}–10^{-9} is supposed as low. Actinomycete strains should be examined for their resistances to antibiotics corresponding to the resistance genes located in the target mobilisable plasmids. Apramycin and thiostrepton resistance genes were commonly used as resistant markers in those plasmids [28, 29, 30, 32, 33].

The spore concentration of *Streptomyces* should be adjusted from 10^3 to 10^8 to obtain discrete colonies on the conjugation plates [23]. For instance, 10^3–10^4 spores concentration of *S. rimosus* was used for mating to avoid patchy lawns on conjugation media that gave some false exconjugants [44]. In the case of using mycelia, concentration of mycelia should also be adjusted. For example, exponential phase mycelium of *S. rimosus* at 10^4 cfu was mated with *E. coli* donor [44].

The ratio of recipient to donor is another factor affected to the efficiency of conjugation. For example, the efficiency of conjugation was increased from 10^{-8} to 10^{-5} exconjugants/recipients when decreasing the donor/recipient ratio for conjugal transfer of pSET152 from *E. coli* ET12567 (pUZ8002) to *S. lavendulae* FRI-5 on ISP 2 medium, but did not affect the frequency of conjugation on MS medium [43]. Similar results were observed in *S. natalensis* [45] and *Kitasatospora setae* (10^{-8} to 10^{-6}) [49] where an excess of donor cells is required to achieve the highest conjugation frequency. However, decreasing the donor/recipient ratio affected the efficiency of conjugation between *E. coli* S17-1 (pOJ436) and *S. toyocaensis* on R2 medium [53]. The conjugation frequency significantly enhanced when number of donor cells of *S. lincolnensis* was increased [55]. In contrast to what occured in *S. coelicolor* and *S. nodosus*, where the number of donor cells per recipient did not affect the efficiency of DNA transfer [23, 39].

After mating on agar medium, incubation period for sufficient growth of actinomycetes at 28–30 ºC is necessary before flooding the agar plates with nalidixic acid and other selective antibiotics. In general, about 16–20 h of incubation is suitable for several streptomycete strains. Rare actinomycetes may take longer incubation time such as endophytic *Micromonospora* sp. GMKU 358 required 24-h incubation before drug overlay.

The incubation temperature also effected the growth of exconjugants in some species. For example, incubation at 37 ºC of conjugation plates between *E. coli* S17-1 and *S. fradiae* gave three times higher efficiency than at 29 ºC [9]. Vector type and size also affected the efficiency of conjugation. In general, the conjugation with small plasmids (approx. 5 kb) has 100–1000 times higher efficiency than that with large plasmids (approx. 20 kb) or cosmids (40–50 kb) [28]. It is advised to test the efficiency of various types of plasmids for effective conjugation for each actinomycete species.

Applications of Intergeneric Conjugation in Actinomycetes

Conjugal transfer by mobilisable plasmids into actinomycete hosts continue to be important in genetic analysis and are routinely used to transfer recombinant plasmids for gene expression or target gene disruption. This genetic transfer system from *E. coli* to actinomycetes enable simple and prospective approach to identify and analyse the function of biosynthetic pathways for medicinally interesting metabolites and precisely manipulate the biosynthesis by pathway engineering of homologous and heterologous genes [61, 66].

Disruption of some target genes through intergeneric conjugation makes verification of the function of a particular gene/gene cluster possible. For instance, disruption of *mycAV* gene in *Micromonospora griseorubida* indicated that it encoded one of the multifunctional mycinamicin PKS genes required during mycinamicin production [59]. When *mycAV* gene was transferred back to the mutant through intergeneric conjugation, complementation was successful [59]. An expression integrative plasmid, pJN100, was used in a complementation experiment of a mutant of *S. griseus* disrupted in the *nonM* gene encoding a reductase essential for nonactin biosynthesis [33]. Disruption of *desD*-like gene in siderophore producing endophytic *Streptomyces* sp. GMKU 3100 could demonstrate the important function of this siderophore trait to plant growth enhancement in rice and mungbean plants [50].

Intergeneric conjugation is used to introduce a gene/gene cluster of interest into the chromosome of an actinomycete host via site specific recombination through actinophage attachment-integrase system located on the integrative plasmids. By this approach, the resistance gene *ptr* copies based on pSET152 was introduced into pristinamycin producing *S. pristinaespiralis* and revealed a dramatic increase in pristinamycin production [40]. New mycinosyl rosamicin-derivatives were obtained when mycinose biosynthesis genes presented in *Micromonospora griseorubida* were introduced via intergeneric conjugation into the rosamicin producing *M. rosaria* IFO13697 [60]. It was demonstrated function of genes responsible for stimulation of NPP biosynthesis of *Pseudonocardia autotrophica* by successfully expression of those genes under the influence of the strong constitutive *ermE** promoter in the integrative expression pSET152 vector [67]. After conjugal transfer system was established for the rimocidin producing *S. rimosus* R7, pIJ8655 containing the *egfp* gene under the control of the thiostrepton-inducible *tipA* promoter was introduced and revealed a high level of fluorescence [44].

Furthermore, a polyester-degrading hydrolase gene from *Thermobifida* sp. BCC23166 was introduced and expressed in *S. rimosus* R7 under *tipA* promoter in pIJ8600 [68]. The biosynthetic gene cluster for microbisporicin, a potent lantibiotic produced by *Microbispora corallina*, was identified by genome scanning [61]. *MibA* gene in the gene cluster was replaced with an apramycin resistance cassette by PCR targeting method [69] and the construct was introduced into *M. corallina* by intergeneric conjugation from ET12567 pUZ8002, and the result indicated that *mibA* gene is essential for microbisporicin biosynthesis [61]. There have been more research works carrying on using intergeneric conjugation procedure to transfer gene/gene cluster and study of gene/gene cluster functions in several actinomycete species.

Conclusion

The development of intergeneric conjugation for genetic exchange system in actinomycetes has become a powerful method to manipulate genes/gene clusters of these pharmaceutically important bacteria. To study the conjugative system of a particular strain, nature of recipients as spores/mycelia has to be examined. Types of *E. coli* donor and mobilisable plasmids must be selected for suitable conjugal transfer with actinomycete species.

Other parameters including the type of conjugation media and concentration of magnesium chloride should be investigated to achieve high conjugation efficiency. Such developed gene transfer system will allow the performance of genetic analysis (e.g. gene expression, target gene disruption) in desired host. The advantages of this genetic transfer system from *E. coli* to actinomycetes allow identification and functional analysis of biosynthetic pathways for pharmaceutically interesting metabolites and the target manipulation of the biosynthesis by pathway engineering of homologous and heterologous genes. Conjugative genetic exchange will provide the means to modify and/or construct strains for novel biotechnological applications.

Acknowledgments

Work from our laboratory related to the material discussed in this chapter was supported by the National Center for Genetic Engineering and Biotechnology (BIOTEC), NSTDA, Thailand Research Fund (TRF), Kasetsart University Research and Development Institute (KURDI) and Center for Advanced Studies in Tropical Natural Resources, National Research University-Kasetsart University (CASTNAR, NRU-KU).

References

[1] Hopwood, D. A. (1999). Forty years of genetics with *Streptomyces*: from *in vivo* through *in vitro* to *in silico*. *Microbiology*, *145*, 2183–2202.

[2] Baltz, R. H. (2008). Renaissance in antibacterial discovery from actinomycetes. *Current Opinion in Pharmacology*, *8*, 557–563.

[3] Watve, M. G., Tickoo, R., Jog, M. M., and Bhole, B. D. (2001). How many antibiotics are produced by the genus *Streptomyces*? *Archives of Microbiology*, *176*, 386–390.

[4] Bibb, M. J., Ward, J. M., and Hopwood, D. A. (1978). Transformation of plasmid DNA into *Streptomyces* at high frequency. *Nature*, *274*, 398–400.

[5] Kieser, T., Bibb, M. J., Buttner, M. J., Chater, K. F., and Hopwood, D. A. (2000). *Practical Streptomyces Genetics*. Norwich, UK: The John Innes Foundation.

[6] MacNeil, D. J. (1988). Characterization of a unique methyl-specific restriction system in *Streptomyces avermitilis*. *Journal of Bacteriology*, *170*, 5607–5612.

[7] Pigac, J., and Schrempf, H (1995). A simple and rapid method of transformation of *Streptomyces rimosus* R6 and other streptomycetes by electroporation. *Applied and Environmental Microbiology*, *61*, 352–356.

[8] Fitzgerald, N. B., English, R. S., Lampel, J. S. and Vanden Boom, T. J. (1998). Sonication-dependent electroporation of the erythromycin producing bacterium *Saccharopolyspora erythraea*. *Applied and Environmental Microbiology*, *64*, 1580–1583.

[9] Mazodier, P., Petter, R., and Thompson, C. (1989). Intergeneric conjugation between *Escherichia coli* and *Streptomyces* species. *Journal of Bacteriology*, *171*, 3583–3585.

[10] Mazodier, P., and Davies, J. (1991). Gene transfer between distantly related bacteria. *Annual Review in Genetics*, *25*, 147–171.

[11] Zechner, E. L., de la Cruz, F., Eisenbrant, R., Grahn, A. M., Koraimann, G., Lanka, E., Muth, G., Pansegrau, W., Thomas, C. M., Wilkins, B. M., and Zatyka, M. (2000). Conjugative-DNA transfer processes. In C. M. Thomas (Ed.), *The Horizontal Gene Pool* (pp. 87–173). London: Harward Academic Press.

[12] Errington, J., Bath, J., and Wu, L. J. (2001). DNA transport in bacteria. *Nature Reviews Molecular Cell Biology*, *2*, 538–544.

[13] Lanka, E., and Wilkins, B. M. (1995). DNA processing reactions in bacterial conjugation. *Annual Reviews in Biochemistry*, *64*, 141–169.

[14] Grohmann, E., Muth, G., and Espinosa, M. (2003). Conjugative plasmid transfer in gram-positive bacteria. *Microbiology and Molecular Biology Reviews*, *67*, 277–301.

[15] Reuther, J., Gekeler, C., Tiffert, Y., Wohlleben, W., and Muth, G. (2006). Unique conjugation mechanism in mycelial streptomycetes: a DNA-binding ATPase translocates unprocessed plasmid DNA at the hyphal tip. *Molecular Microbiology, 61*, 436–446.

[16] Possoz, C., Ribrad, C., Gagnat, J., Pernodet, J.-L., and Guérlineau, M. (2001). The integrative element pSAM2 from *Streptomyces*: Kinetics and mode of conjugal transfer. *Molecular Microbiology, 42*, 159–166.

[17] Kosono, S., Kataoka, M., Seki, T., and Yoshida, T. (1996). The TraB protein, which mediates the intermycelial transfer of the *Streptomyces* plasmid pSN22, has functional NTP-binding motifs and is localized to the cytoplasmic membrane. *Molecular Microbiology, 19*, 397–405.

[18] Kieser, T., Hopwood, D. A., Wright, H. M., and Thompson, C. J. (1982). pIJ101, a multicopy broad host-range *Streptomyces* plasmid: Functional analysis and development of DNA cloning vectors. *Molecular and General Genetics, 185*, 223–238.

[19] Hagege, J., Pernodet, J. L., Friedmann, A., and Guerineau, M. (1993). Mode and origin of replication of pSAM2, a conjugative integrating element of *Streptomyces ambofaciens*. *Molecular Microbiology, 10*, 799–812.

[20] Bibb, M. J., Freeman, R. F., and Hopwood, D. A. (1977). Physical and genetic characterization of a second sex factor, Scp2, for *Streptomyces coelicolor* A3(2). *Molecular and General Genetics, 154*, 155–166.

[21] Hopwood, D. A. and Kieser, T. (1993). Conjugative plasmids of *Streptomyces*. In D. B. Clewell (Ed.), *Bacterial Conjugation* (pp. 293–311). New York: Plenum Press.

[22] Trieu-Cuot, P., Carlier, C., Martin, P., and Courvalin, P. (1987). Plasmid transfer by conjugation from *E. coli* to Gram-positive bacteria. *FEMS Microbiology Letters, 48*, 289–294.

[23] Flett, F., Mersinias, V., and Smith, C. P. (1997). High efficiency intergeneric conjugal transfer of plasmid DNA from *Escherichia coli* to methyl DNA-restricting streptomycetes. *FEMS Microbiology Letters, 135*, 223–229.

[24] MacNeil, D. J., Gewain, K. M., Ruby, C. L., Dezeny, G., Gibbons, P. H., and MacNeil, T. (1992). Analysis of *Streptomyces avermitilis* genes required for avermectin biosynthesis utilizing a novel integration vector. *Gene, 111*, 61–68.

[25] Sia, L. A., Kuehner, D. M., and Figurski, D. H. (1996). Mechanism of retrotransfer in conjugation: Prior transfer of the conjugative plasmid is required. *Journal of Bacteriology, 178*, 1457–1464.

[26] Paget, M. S. B., Chamberlin, L., Atrih, A., Foster, S. J., and Buttner, M. J. (1999). Evidence that the extracytoplasmic function sigma factor σ^E is required for normal cell wall structure in *Streptomyces coelicolor* A3(2). *Journal of Bacteriology, 181*, 204–211.

[27] Kuhstoss, S., and Rao, R. N. (1991). Analysis of the integration function of the streptomycete bacteriophage ϕC31. *Journal of Molecular Biology, 222*, 897–908.

[28] Bierman, M., Logan, R., O'Brien, K., Seno, E. T., Nagaraj, R., and Schoner, B. E. (1992). Plasmid cloning vectors for the conjugal transfer of DNA from *Escherichia coli* to *Streptomyces* spp. *Gene, 116*, 43–49.

[29] Sun, J., Kelemen, G. H., Fernández-Abalos, J. M., and Bibb, M. J. (1999). Green fluorescent protein as a reporter for spatial and temporal gene expression in *Streptomyces coelicolor* A3(2). *Microbiology, 145*, 2221–2227.

[30] Huang, J., Shi, J., Molle, V., Sohlberg, B., Weaver, D., Bibb, M. J., Karoonuthaisiri, N., Lih, C.-J., Kao, C.M., Buttner, M. J., and Cohen, S. N. (2005). Cross-regulation among disparate antibiotic biosynthesis pathways of *Streptomyces coelicolor. Molecular Microbiology, 58,* 1276–1287.

[31] Hong, H.-J., Hutchings, M. I., Hill, L. M., and Buttner, M.J. (2005). The role of the novel fem protein Vank in vancomycin resistance in *Streptomyces coelicolor. Journal of Biological Chemistry, 280,* 13055–13061.

[32] Hillemann, D., Puhler, A., and Wohlleben, W. (1991). Gene disruption and gene replacement in *Streptomyces* via single stranded DNA transformation of integration vectors. *Nucleic Acids Research, 19,* 727–731.

[33] Nikodinovic, J., and Priestley, N. D. (2006). A second generation *snp*-derived *Escherichia coli – Streptomyces* shuttle expression vector that is generally transferable by conjugation. *Plasmid, 56,* 223–227.

[34] Voeykova, T., Emelyanova, L., Tabakov, V., and Mkrtumyan, N. (1998). Transfer of plasmid pOT101 from *Escherichia coli* to various representatives of the order *Actinomycetales* by intergeneric conjugation. *FEMS Microbiology Letters, 162,* 47–52.

[35] Voeykova, T. A., Emelyanova, L. K., and Kudryavtseva, E. A. (2003). Development of the system of plasmid transfer to strains *Streptomyces avermitilis* ATCC 31272 and *Saccharopolyspora erythraea* ATCC 11635 by the method of intergeneric conjugation. *Russian Journal of Genetics, 39,* 543–552.

[36] Fouces, R., Rodriguez, M., Mellado, E., Diez, B., and Barredo, J. L. (2000). Conjugation and transformation of *Streptomyces* species by tylosin resistance. *FEMS Microbiology Letters,* 186, 319–325.

[37] Blaesing, F., Műhlenweg, A., Vierling, S., Ziegelin, G., Pelzer, S., and Lanka, E. (2005). Introduction of DNA into actinomycetes by bacterial conjugation from *E. coli* – An evaluation of various transfer systems. *Journal of Biotechnology, 120,* 146–161.

[38] Guan, D., and Pettis, G. S. (2009). Intergeneric conjugal gene transfer from *Escherichia coli* to the sweet potato pathogen *Streptomyces ipomoeae. Letters in Applied Microbiology, 49,* 67–72.

[39] Nikodinovic, J., Barrow, K. D., and Chuck, J.-A. (2003). High frequency transformation of the amphotericin-producing bacterium *Streptomyces nodosus. Journal of Microbiological Methods, 55,* 273–277.

[40] Jin, Z., Jin, X., and Jin, Q. (2010). Conjugal transferring of resistance gene *ptr* for improvement of pristinamycin-producing *Streptomyces pristinaespiralis. Applied Biochemistry and Biotechnology, 160,* 1853–1864.

[41] Myronovskyy, M., Ostash, B., Ostash, I., and Fedorenko, V. (2009). A Gene cloning system for the siomycin producer *Streptomyces sioyaensis* NRRL-B5408. *Folia Microbiologica, 54,* 91–96.

[42] Hou Y.-H., Li, F.-C., Wang S.-J., Qin S., and Wang, Q.-F. (2008). Intergeneric conjugation in holomycin-producing marine *Streptomyces* sp. strain M095. *Microbiological Research, 163,* 96–104.

[43] Kitani, S., Bibb, M. J., Nihira, T., and Yamada, Y. (2000). Conjugal transfer of plasmid DNA from *Escherichia coli* to *Streptomyces lavendulae* FRI-5. *Journal of Microbiology and Biotechnology,* 10, 535–538.

[44] Phornphisutthimas, S., Sudtachat, N., Bunyoo, C., Chotewutmontri, P., Panijpan, B., and Thamchaipenet, A. (2010). Development of an intergeneric conjugal transfer

system for rimocidin-producing *Streptomyces rimosus*. *Letters in Applied Microbiology*, *50*, 530–536.

[45] Enŕiquez, L. L., Mendes, M. V., Antón, N., Tunca, S., Guerra, S. M., Martín, J. F., and Aparicio, J. F. (2006). An efficient gene transfer system for the pimaricin producer *Streptomyces natalensis*. *FEMS Microbiology Letters*, *257*, 312–318.

[46] Nikodinovic, J., Barrowa, K. D., and Chuck, J.-A. (2003). High frequency transformation of the amphotericin-producing bacterium *Streptomyces nodosus*. *Journal of Microbiological Methods*, *55*, 273–277.

[47] Huiqun, D., Xiaofeng, C., Jianxin, P., Huazhu, H., Koji, I., and Aiying, I. (2010). Practical procedures for genetic manipulation systems for medermycin-producing *Streptomyces* sp. AM-7161. *Journal of Basic Microbiology*, *50*, 299–301.

[48] Ha, H.-S., Hwang, Y.-I. and Choi, S.-U. (2008). Application of conjugation using ϕC31 att/int system for *Actinoplanes teichomyceticus*, a producer of teicoplanin. *Biotechnology Letters*, *30*, 1233–1238.

[49] Choi, S.-U., Lee, C.-K., Hwang, Y.-H., Kinoshita, H., and Nihira, T. (2004). Intergeneric conjugal transfer of plasmid DNA from *Escherichia coli* to *Kitasatospora setae*, a bafilomycin B1 producer. *Archive in Microbiology*, *181*, 294–298.

[50] Rungin, S., Indananda, C., Suttiviriya, P., Kruasuwan, W., Jaemsaeng, R., and Thamchaipenet, A. (2012). Plant growth enhancing effects by a siderophore-producing endophytic streptomycete isolated from a Thai jasmine rice plant (*Oryza sativa* L. cv. KDML105). *Antonie van Leeuwenhoek*, *102*, 463–472.

[51] Kim, M.-K., Ha, H.-S., and Choi, S.-U. (2008). Conjugal transfer using the bacteriophage ϕC31 *att*/*int* system and properties of the *attB* site in *Streptomyces ambofaciens*. *Biotechnology Letters*, *30*, 695–699.

[52] Hamano, Y., Nicchu, I., Hoshino, Y., Kawai, T., Nakamori, S., and Takagi, H. (2005). Development of gene delivery systems for the ε-poly-L-lysine producer, *Streptomyces albulus*. *Journal of Bioscience and Bioengineering*, *99*, 636–641.

[53] Matsushima, P., and Baltz, R. H. (1996). A gene cloning system for '*Streptomyces toyocaensis*'. *Microbiology*, *142*, 261–267.

[54] Paranthaman, S. and Dharmalingum, K. (2003). Intergeneric conjugation in *Streptomyces peucetius* and *Streptomyces* sp. strain C5: chromosomal integration and expression of recombinant plasmids carrying the *chi*C gene. *Applied and Environmental Microbiology*, *60*, 84–91.

[55] Du, L., Liu, R.-H., Ying, L., and Zhao, G.-R. (2012). An efficient intergeneric conjugation of DNA from *Escherichia coli* to mycelia of the lincomycin-producer *Streptomyces lincolnensis*. *International Journal of Molecular Science*, *13*, 4797–4806.

[56] Stinchi, S., Azimonti, S., Donadio, S., and Sosio, M. (2003). A gene transfer system for the glycopeptides producer *Nonomuraea* sp. ATCC39727. *FEMS Microbiology Letters*, *225*, 53–57.

[57] Li, X., Zhou, X., and Deng, Z. (2003). Vector systems allowing efficient autonomous or integrative gene cloning in *Micromonospora* sp. strain 40027. *Applied and Environmental Microbiology*, *69*, 3144–3151.

[58] Rose, K., and Steinbüchel, A. (2002). Construction and intergeneric conjugative transfer of a pSG5-based cosmid vector from *Escherichia coli* to the polyisoprene

rubber degrading strain *Micromonospora aurantiaca* W2b. *FEMS Microbiology Letters*, 211, 129–132.

[59] Anzai, Y., Ishii, Y., Yoda, Y., Kinoshita, K., and Kato, F. (2004). The targeted inactivation of polyketide synthase *mycAV* in the mycinamicin producer, *Micromonospora griseorubida*, and a complementation study. *FEMS Microbiology Letters, 238*, 315–320.

[60] Anzai, Y., Iizaka, Y., Li, W., Idemoto, N., Tsukada, S., Koike, K., Kinoshita, K., and Kato, F. (2009). Production of rosamicin derivatives in *Micromonospora rosaria* by introduction of D-mycinose biosynthetic gene with ϕC31-derived integration vector pSET152. *Journal of Industrial Microbiology and Biotechnology, 36*, 1013–1021.

[61] Foulston, L. C., and Bibb, M. J. (2010). Microbisporicin gene cluster reveals unusual features of lantibiotic biosynthesis in actinomycetes. *Proceeding of the National Academy Science of the USA, 107*, 13461–13466.

[62] Marcone, L. G, Foulston, L., Binda, E., Marinelli, F., Bibb, M., and Beltrametti, F. (2010). Methods for the genetic manipulation of *Nonomuraea* sp. ATCC 39727. *Journal of Industrial Microbiology and Biotechnology*, 37, 1097–1103.

[63] Bradley, D. E., Taylor, D. E., and Cohen, D. R. (1980). Specification of surface mating systems among conjugative drug resistance plasmids in *Escherichia coli* K-12. *Journal of Bacteriology, 143*, 1466–1470.

[64] Hobbs, G., Frazer, C. M., Gardner, D. C. J., Cullum, J. A. and Oliver, S. G. (1989). Dispersed growth of *Streptomyces* in liquid culture. *Applied Microbiology and Biotechnology, 31*, 272–277.

[65] Shirling, E. B. and Gottlieb, D. (1966). Method of characterization of *Streptomyces* species. *International Journal of Systematic Bacteriology, 16*, 3313–3340.

[66] Courvalin, P. (1994). Transfer of antibiotic resistance gene between Gram-positive and Gram-negative bacteria. *Antimicrobial Agents and Chemotherapy, 38*, 1447–1451.

[67] Jeon, H.-G., Seo, J., Lee, M.-J., Han, K., and Kim, E.-S. (2011). Analysis and functional expression of NPP pathway-specific regulatory genes in *Pseudonocardia autotrophica*. *Journal of Industrial Microbiology and Biotechnology, 38*, 573–579.

[68] Sinsereekul, N., Wangkam, T., Thamchaipenet, A., Srikhirin, T., Eurwilaichitr, L., and Champreda, V. (2010). Recombinant expression of BTA hydrolase in *Streptomyces rimosus* and catalytic analysis on polyesters by surface plasmon resonance. *Applied Microbiology and Biotechnology, 86*, 1775–1784.

[69] Gust, B., Challis, G. L., Fowler, K., Kieser, T., Chater, K. F. (2003). PCR-targeted *Streptomyces* gene replacement identifies a protein domain needed for biosynthesis of the sesquiterpene soil odor geosmin. *Proceeding of the National Academy Science of the USA, 100*, 1541–1546.

In: Molecular Biology of Bacteria
Editors: R. Kumar Gaur and H. K. Gautam
ISBN: 978-1-62618-189-2
© 2013 Nova Science Publishers, Inc.

Chapter 2

Metagenomics: A Hunting Expedition in Microbial Diversity

Rajal Debnath, Rupak K. Sarma, Ratul Saikia*, Archana Yadav and Tarun C. Bora
Biotechnology Division,
CSIR- North East Institute of Science and Technology,
Jorhat, Assam, India

Abstract

Many inhabitants of the microscopic world escape our curious eyes because of their reluctance to grow in pure culture. The cultured community of microbes represent only 0.1% to 1% of the organisms in most habitats. Metagenomics intends to explore the rest of 99% of the population which remains unnoticed. This culture-independent approach targets the total community genetic material and their subsequent cloning. Sequence-based approach or function-based screening is used to discover the positive clones for viable noble activity, etc. by combining high throughput screens and robust computational algorithms.

The gigabytes of data generated in the process helps in relative genome sequencing projects, metabolomics, meta-proteomics and meta-transcriptomics studies. This chapter outlines the readers, the various tools and methodologies developed and the trends that follow in metagenomics study along with recent technological advances. Metagenomics will continue to be a powerful tool to understand the community, function between microbial populations in their native environment thereby developing biotechnological applications.

* Corresponding author: Email: rsaikia19@gmail.com; saikia_ratul@rrljorhat.res.in; Phone: +91-376-237295; Fax: +91-376- 2370011.

1. Introduction

Inception of microbial genomes study started in the late 1970s, the genomes of bacteriophages MS2 (Fiers et al. 1976) and QX174 (Sanger et al. 1978) were the first to enter the sequencing projects. In 1995 microbiologists undertook a major step with the sequencing of the first bacterial genome *Haemophilus influenza* (Fleischmann et al. 1995). With the accumulation of large number of microbial genomes data, microbial evolution studies have greatly changed. GenBank release 2.2.6 depicted 916 bacterial genomes, 1,987 viral and 67 archaeal species. However, there are limits in studying single organism genomes. First, technological limitation is that an organism must first be clonally cultured to sequence its entire genome. Since only a small percentage of microbes in nature can be cultured, extant genomic data are highly biased and do not represent a true picture of the genomes of microbial species (Amann et al. 1995, Pace et al. 1997, Rappe et al. 2003). With the advent of economic and high throughput sequencing technologies a lot of these hurdles are overcome. We now have the ability to obtain genomic information directly from microbial communities in their natural habitats. Moreover, instead of looking at a few species individually, we are able to study tens of thousands all together. Sequence data taken directly from the environment denotes metagenome (Handelsman et al. 1998), and the study of these sequence data directly from the environment encompasses metagenomics (Rondon et al. 2000). It is a collective study of genomes recovered from environmental samples without prior cultivation. It is therefore a means of systematically investigating, classifying and manipulating the entire genetic material isolated from environmental samples. Taxonomy based on phylogeny, in which data from uncultured bacteria are included, is rapidly changing and replacing the former taxonomy based exclusively on morphological, physiological and biochemical parameters of cultured bacteria (Garrity et al. 2005). The idea of cloning DNA directly from environmental samples was first proposed by Pace et al. (1985) and in 1991, the first such cloning in a phage vector was reported (Schmidt et al. 1991). The approach to metagenomic library screens for novel enzymes or drug discoveries utilizes the direct extraction of total genetic materials from environmental samples without the need for culturing. The genetic material is then ligated into vectors and transferred into new host cells (mainly various competent cells of *Escherichia coli*) to generate metagenomic libraries. These libraries can then be used to study structural and functional diversity by DNA sequencing or to search for new products based on sequence mining or functional expression. It is important, however, that the samples for metagenomics should be targeted from particular sites to increase the success rate because metabolites discovered by this approach depend directly on microorganisms present in the samples.

2. Sampling and Generation of Metadata

The problem in microbial ecology is that we are unable to see how many samples are enough to comprise the entire population that is sought to study. Rarefaction curves are typically used to estimate the fraction of species sequenced. It is a plot of number of species versus number of individuals sampled, which typically defines the population richness and evenness. ''Data about the data'' is what Metadata refers to: where the samples were taken from, when, and under which conditions. Generation and recording of metadata is as important as the sequence data. In microbial ecology, this commonly refers to physical, chemical, and other environmental characteristics of the sample's location. For an instance, an ocean sample metadata will typically include sampling date and time, depth, salinity, light intensity, geographical coordinates, pH, soluble gases, etc. whereas in clinical microbiology, metadata would refer to the pathology, medical history, and vital statistics of the patient as well as the exact location and tissue from which the sample was taken, the sampling conditions, and so on. Many metagenomic studies are driven by discovery and data mining, rather than by hypothesis. These studies seek statistically significant correlations between the metagenomic data and the habitat-associated metadata, which may lead to biologically significant discoveries. There is therefore a need to provide metadata in a form that is standard, comprehensive, and amenable to computation. In a recent publication, a standard for the Minimum Information of Genomic and Metagenomic (MIGS/MIMS) metadata was suggested for adoption (Field et al. 2008), and an associated mark up language, the Genomics Contextual Data Markup Language (GCDML) is under active development (Kottmann et al. 2008). It is the consortium's aim that the MIGS/MIMS shall be adopted by journals as a publication requirement when genomic or metagenomic data are being deposited.

3. Isolation and Enriching the Environmental DNA

A number of methods on extraction of metagenomic DNA have been published (Picard et al. 1992, Smalla et al. 1993, Holben et al. 1994, Zhou et al. 1996, Burgmann et al. 2001, Kauffmann et al. 2004, and Bertrand et al. 2005). In all the described protocols, presence of humic acids as low as 10 ng, which is a major inhibitor of PCR reactions has posed difficulty in downstream applications of the isolated DNA. These polyphenols are also known to interfere with enzymatic modifications of purified DNA. Polyvinylpolypyrrolidone (PVPP), gel purification, density gradient centrifugation, CTAB, size exclusion spin column were used in metagenomic protocols to remove such inhibitors (Tsai et al. 1992). Humic acid and fulvic acid along with other organic pollutants, metal ions and chemical impurities co-precipitate along with the metagenomic DNA and interfere with the subsequent diversity analysis (Hinoue et al. 2004). Also inert materials like amberlite resins and powdered activated charcoals (PAC) are efficient removers of metal ions, coloured compounds etc. (Desai et al. 2006). PACs due to their large surface area and pore volume are used to adsorb and remove lignin sulfonate, tannic acid, arabic gum and other biodegradable organic substances whereas amberlite resins as ion exchangers remove heavy metals like Cu (II), Pb(II), Zn(II), Ni(II) and Cr(VI).

High-quality DNA that has not been fragmented is needed which requires less vigorous methods than those applied to diversity studies (e.g. bead beating). High-molecular weight DNA is desirable because biosynthetic gene clusters encoding antimicrobial and other compounds of industrial interest typically range from 30 to 100 Kb. Thus, the recovery of high-molecular-weight DNA and subsequent insertion of large inserts is the most efficient method for identifying new enzymes.

In enrichment culture approaches, isolated environmental DNA can be subjected to whole genome amplification that is multiple displacement amplification (MDA) to provide sufficient genetic substrate for library production. In some instances, such as low-nutrient environments (e.g. recent deglacial areas and rock), it might be problematic to obtain sufficient DNA for cloning. For such cases, whole-genome amplification can be used. Although this method is based on the use of short oligonucleotides that randomly amplify genomic DNA, it remains subject to the potential biases of PCR-based approaches (Bodelier et al. 2009). Multiple displacement amplification uses Ø29 DNA polymerase and a constant incubation temperature (Abulencia et al. 2006) to increase the amount of template DNA. Although multiple displacement amplification products are generally larger than those achieved from conventional whole-genome amplification approaches, they remain subject to amplification bias and the creation of chimeric sequences. It remains clear that better extraction efficiencies need to be developed for the recovery of good-quality DNA from difficult samples (e.g. highly humic soils or low template concentration), and some novel extraction methods are being explored (Yun et al. 2004).

There are other methodologies to enrich environmental DNA like differential expression analysis (DEA) which employs the principle of transcriptional differences in gene expression and detection of target genes, 5-Bromo-2-deoxyuridine (BrdU) labelling which is isolation based on BrdU-labeled DNA in metabolically active cells. In stable isotope probing (SIP) microbial DNA/RNA is probed with stable isotopes followed by separation of labeled/unlabeled DNA by ultracentrifugation. Suppressive subtractive hybridization (SSH) is another enrichment technique where subtractive hybridization is carried out to select DNA fragments unique to each DNA sample and identification of genetic differences between microorganisms. Likewise microarray method and phage display expression system can also be employed to enrich DNA.

4. Metagenomic Nucleic Acid Isolation Kits

Commercial kits are also available for the extraction and purification of soil metagenomic DNA (Rajendhran et al. 2008). These kits take lesser processing time and yield inhibitor free metagenomic DNA form soil and sediment samples. UltraClean™, PowerSoil™, PowerMax™ (Mo Bio Laboratories Inc., Carlsbad, CA, USA), SoilMaster™ (Epicentre Biotechnologies, Madison, WI, USA) and FastDNA® Spin kit for soil (MP Biomedicals, Solon, Oh, USA) are the commonly used kits for soil metagenomic DNA isolation. WaterMaster™ DNA purification kit from (Epicentre Biotechnologies, Madison, WI, USA) efficiently isolates total DNA from water samples.

Recently, RNA PowerSoil™ (Mo Bio) and FastRNA® Pro soil-direct kit (MP Biomedicals) were introduced for the isolation of soil metagenomic RNA. In UltraClean™ and PowerMax™ kits, cells are lysed by mechanical (bead beating) and chemical methods and the extracted DNA is purified using silica-based spin column.

In PowerMax™, an additional humic substance/brown colour removal procedure is included for the efficient removal of PCR inhibitors (www.mobio.com). SoilMaster™ kit utilizes a hot detergent method for cell lysis and spin column for purification (www.epibio.com). FastDNA® Spin kit for soil is designed with a dedicated FastPrep® instrument.

This bench-top device is used for the mechanical lysis of cells by homogenization with a mixture of ceramic and silica particles and the extracted DNA is purified with spin column (www.mpbio.com). Puregene DNA purification kit, QIAamp® stool mini kit from Qiagen (Hilden, Germany) and Wizard™DNA clean-up kit from Promega (Madison, WI, USA) were also used for the isolation of soil metagenomic DNA with modified protocols.

5. Function-Based Metagenomic Library Screening

The goals of metagenomic projects may be the - i) identification of functional genes, ii) estimation of the microbial diversity, iii) understanding the population dynamics of a whole community or iv) assembly of complete genome of an uncultured organism. *E. coli* are the most efficient host system for the creation of libraries in metagenomic studies. The reason being the absence of recombination (*recA, recBC*) and restriction (*mcrA, mcrBC*) genes. Highly efficient genetically modified strains of *E coli* competent cells are commercially available. Plasmids are suitable for cloning smaller inserts of 10-kb DNA fragments, whereas cosmids (25–35 kb), fosmids (25–40 kb), or BACs (100–200 kb) can be used to clone larger fragments. Enzyme activities are usually assayed on agar plates supplemented with substrates. Identification of a positive clone through visual screening can be done by culturing the metagenomic library on plates and observation of clear halo or colour. Another strategy to improve sensitivity is to use cell lysates for screening wherein chemically or physically prepared lysates from the library of cells are grown on 96-well plates and mixing them with the substrates. In reporter assay based approach, expression of the more sensitive reporter gene *viz.*, β-galactosidase, green fluorescent protein (GFP) and tetracycline resistance gene are conducted after linking them to target biological events (Guan et al. 2007, Uchiyama et al. 2005, Riaz et al. 2008). Currently, *E. coli* is the dominant screening host for functional metagenomics and is utilized as a replication host by most commercially available large insert library production systems. Indeed, numerous *E. coli* strains have been engineered that allow for stable replication of various single-copy or multi-copy vectors, minimal recombination and protection from lytic phage. *E. coli* harbouring fosmids or BACs promotes stable replication and minimizes toxic genes overexpression due to their low copy number status within. However, the same attributes that make large insert clones useful for genomic sequencing can become liabilities when searching for function within a metagenomic library. Some of the common expression host used in metagenomic screens are *Agrobacterium tumefaciens, Caulobacter vibroides, Rhizobium leguminosarum, Burkholderia graminis, Ralstonia metallidurans, Escherichia coli, Pseudomonas fluorescens, Pseudomonas putida, Xanthomonas campestris* etc. These strains are all gram negative and the phenotypic trait used in screens are either based on pigmentation, morphology, antibiosis or enzymatic activity like alcohol/ aldehyde dehydrogenase. This form of visible screen has drawbacks of its own. Visible screens are labour-intensive but require no special equipment beyond careful observation. However, signal-to-noise associated with colony detection is low, and weakly expressing positive clones are often missed.

Developing effective screens based on genetic selection and survival of the host is the most powerful due to their digital nature of output either as growth or no growth. Complementation of genes mutated or missing in the host is so far the significant approach for screens. Additional innovative screening approaches such as Substrate - Induced Gene Expression screening (SIGEX) have facilitated isolation of novel catabolic genes from environmental metagenomes, particularly genes that are difficult to obtain using conventional gene-cloning methods. Promoter trap vectors are used in this method owing to their great promise for high throughput discovery. In this technique promoter less green fluorescent protein (GFP) expression vector for use in environmental DNA cloning. Substrate-induced gene expression of the resulting clones enabled the recovery of GFP positive clones grown in liquid culture using fluorescence-activated cell sorting (FACS). Similarly, detection of enzyme activity using colorimetric or fluorometric substrates allows for optical detection of positive clones, acquiescent to high throughput methods including automation. Combination of optical detection approaches with flow cytometry and cell sorting can further increase screening throughput.

5. Sequence-Based Metagenomic Library Screening

Searching for novel enzymes from metagenomic libraries can be adopted by sequence based approach which utilizes the use of a set of degenerate oligonucleotide primers that are designed based on the consensus amino acid sequences from the target protein (Morimoto et al. 2009, Tang et al.2008, Wang et al. 2008). This method targets part of the internal gene sequence, and in order to obtain the entire full length gene one needs to clone the flanking regions. The use of metagenome library as a template in such a screen circumvents the need for tedious gene walking steps.

Isolation of novel proteins in this approach is based on homology searches. Although it impedes the discovery of entirely new gene sequences, it is an efficient method of isolating enzymes with different levels of similarity from previously identified genes. For instance, sequence-based screening from the metagenome of a hindgut microbiota of a wood-feeding higher termite identified 700 domains with homology to the glycoside-hydrolase catalytic site (Wang et al. 2008). Recently novel methods such as pre-amplification inverse-PCR (PAI PCR) and metagenomic DNA shuffling have been employed to isolate new biocatalysts. Difficulty in cloning potential genes from low copy number target DNA sequences was overcome by using PAI-PCR which was employed to isolate glycosyl hydrolase from gut microbiota of horse and termite. Spencer et al. (2003) showed SNP (single nucleotide polymorphism) analysis at high throughput scale among the pseudomonas metagenome reads (Spencer et al. 2003).

6. Microbial Diversity Analysis of the Metagenome

To draw ecological inferences from metagenomic data, complete sequencing of clones containing phylogenetic anchors, such as the 16S rRNA gene and the archaeal DNA repair gene *radA* (Beja et al. 2002, Quaiser et al. 2003, Schleper et al. 1998) are carried out, that indicate the taxonomic group and functional information about the organisms from which these clones were derived. The significance of sequence-based screening is its independence from the expression of cloned genes by foreign hosts. Lane et al. (1985) pioneered the metagenomics based microbial diversity analysis and suggested that the direct analysis of 16S rRNA gene sequences in the environment can be used to study the diversity. Initially, sequencing of 16S rRNAs were performed using reverse transcriptase. However, formation of secondary structures in rRNA sequences hampered the sequencing of full length genes. The validity of metagenomic approach in microbial diversity analysis depends on obtaining representative nucleic acids from entire microbial communities. Incomplete cell lysis, DNA adsorption to soil surfaces, co-extraction of enzymatic inhibitors and degradation of DNA at various steps of extraction procedures may hinder the microbial diversity analysis. Robust and high throughput technologies such as Roche / 454 / Pyro / FLX, Illumina/Solexa Genome Analyzer, Applied Biosystems SOLiD^{TM} Sequencer which are rapid parallel sequencing method have greatly hastened the diversity studies of microbes from distantly complex environmental niches. The community structure of diverse groups of eubacteria, archaea has also exploited the quorum sensing ability which bacteria use to coordinate certain behaviors based on the local density of their population.

Microbial community exploration from surface waters of the Sargasso Sea (Venter et al. 2004), using different phylogenetic biomarker genes to elucidate the composition of the bacterial community, identified nine major bacterial phyla viz., (Proteobacteria, Actinobacteria, Cyanobacteria, Firmicutes, Bacteroidetes, Chloroflexi, Spirochaetes, Fusobacteria and Deinococcus-Thermus) and two archaeal phyla (Crenarchaeota and Euryarchaeota). Over 1800 bacterial species, including 148 new phylotypes, were detected, with the vast majority of all sequences clustering within the α- and γ-divisions of the phylum Proteobacteria.

7. The Human Gut Metagenome

The vast majority of our microbial partners live in the intestine. They perform functions such as extracting nutrients and calories from otherwise indigestible components of our diets and making essential vitamins and amino acids. Some gut bacteria also help detoxify potentially harmful chemicals contained in what we eat. Some of the microbes living in and on the human body play a critical role in defending us against pathogens. This mutually beneficial relationship helps protect us from getting diseases while giving the microbes a place to live.

To understand and exploit the impact of the gut microbes on human health and well-being it is necessary to decipher the content, diversity and functioning of the microbial gut community. It has been estimated that the microbes in our bodies collectively make up to 100 trillion cells, tenfold the number of human cells, and suggested that they encode 100-fold more unique genes than our own genome. 16S ribosomal RNA gene (rRNA) sequence-based methods revealed that two bacterial divisions, the Bacteroidetes and the Firmicutes, constitute over 90% of the known phylogenetic categories and dominate the distal gut microbiota. Some of the disease related microbiome studies carried out at JCVI and collaborative institutes are progression of esophageal cancer, bacterial vaginosis and pre-term delivery, nasopharynx microbiome and vaccination in children, skin microbiome, acne and psoriasis, oral diseases including periodontitis, Type 1 Diabetes pilot, astronaut microbiome, ferret respiratory tract infections as an influenza model, spinal cord injuries and urinary tract infections etc.

8. Bioinformatics Tools and Databases for Metagenomics

Computational biology aspect of metagenomics to handle large pile of data owing to the robustness of next generation sequencing technologies has led to the development of tools and software's simultaneously. Algorithms like AMPHORA (http://bobcat.genomecenter.ucdavis.edu/amphora) are used for phylogenetic analysis of single gene or whole genomes, whereas ARB is an interacting software tool for sequence database (RNA/DNA/amino acids) (http://www.arb-home.de) maintenance and analysis. Dedicated metagenomic database such as CAMERA (http://camera.calit2.net) which holds data from collective projects like Sorcerer II Global Ocean Sampling Expedition (Yooseph et al. 2007), Hawaii Ocean Time Series Station ALOHA (Angly et al. 2006), Marine Viromes (Pruitt et al. 2005) and Non redundant sequence database. GOLD (http://www.genomesonline.org) is another portal where metadata information related to genome and metagenome projects worldwide are stored. Megx.net (http://www.megx.net) harbours database tools for analysis of marine metagenomics. RAST (http://metagenomics.nmpdr.org) is used for automatic phylogenetic and functional analysis of genome. Likewise UniFrac and XplorSeq are used for microbial community structure study using phylogenetic information and compilation, management and phylogenetic analysis of DNA sequences respectively. NCBI's RefSeq (http://www.ncbi.nlm.nih.gov/RefSeq) database is a non-redundant collection of sequences representing genomic data, transcripts and proteins from 2400 organisms.

9. Mobile Metagenomics of Plasmid, Transposable Elements and Virus

Genes present on mobile genetic elements (MGE) that populate soil ecosystems constitutes the mobilome of the environment (Rajendhran et al. 2008). MGE include plasmids, transposons, insertion sequences and integrons, which may move between bacterial cells in a population or mobilize into a new host species and introduce new genetic material. The horizontal gene transfer between species is a driving factor for bacterial adaptation and evolution. Several strategies exist for the isolation of plasmids from communities. Treatment with a novel exonuclease, "plasmid-safe DNase", is widely used to selectively remove linear DNA fragments. Plasmid-safe DNase selectively removes double-stranded and single-stranded linear DNA, also single-stranded circular DNA without affecting closed circular supercoiled or nicked circular double-stranded DNA molecules, thus effectively enriching the community plasmid DNA. Such enriched plasmid DNA may be used for shot-gun library creation and sequenced to explore the plasmid encoded genes. In transposon-aided capture (TRACA) system plasmids independent of selectable markers or replication origin are acquired from metagenomics DNA extracts. Instead, all plasmids in the enriched population are provided with an origin of replication and a selectable marker, by tagging plasmids with a transposon encoding these functions in an *in vitro* reaction. Use of transposons with a greater range of host organisms such as Pseudomonas and Streptomyces, replication origins and selectable markers may improve the efficiency of this technique for the isolation of plasmids from various groups of microorganisms.

Due to their integrative nature transposable elements, integrons and genomic islands are also considered in studying mobile metagenomics and they can be readily detected in sequence data from standard metagenomic libraries using computational biology. Bacteriophages with the ability of transferring genetic material between bacterial hosts can also be considered in mobilome studies. Bacteriophage particles are first separated from the environmental samples and nucleic acids are then extracted. Interference due to free and cellular DNA in the environment has posed difficulty in isolation viral metagenome. A combination of differential filtration with tangential flow filters, DNase treatment and density gradient centrifugation in caesium chloride is used for separation of intact viral particles from rest of the metagenome (Rajendhran et al. 2008). Due to modified bases in phage genome, cloning of phage metagenomic DNA in *E. coli* host is cumbersome. In order to replace these modified bases in the phage genome so as to make it amenable for library creation PCR amplification or whole genome amplification can be carried out.

Conclusion

With the massive data generated from the Sargasso Sea project, it is quite vivid that the microbial diversity underlying in vastly different geographical niches holds potential in answering key aspects of life processes ranging from microscopic to macroscopic forms. Metagenomics in this regard has grasped the attention of microbial ecologist around the globe to exploit this microscopic world right from exotic to familiar habitats including deep sea thermal vents; acidic hot springs; arid cold soils; frozen lakes and glaciers; and eukaryotic host organs—the human skin oral and gut microflora, termite and caterpillar guts, plant rhizospheres and phyllospheres. It is however an arduous task to take a particular approach in metagenomics study, because the protocol adopted depends on the target set. In essence the genomic libraries generated using metagenomic approaches would provide firm hold on the conservation and the sustainable maintenance of genetic resources, which could be harnessed in the long run for our own survival owing to the threats posed by global warming and depleting food resources, and improving health care and life style.

Acknowledgments

This study is supported by a Network Project (AMAAS), sponsored by the Indian Council of Agricultural Research (ICAR), Government of India, New Delhi. The authors are thankful to the funding agency. The authors are also thankful to the Director, CSIR-NEIST, Jorhat, India for providing necessary facilities to carry out the work.

References

Abulencia, C.B. et al. (2006). Environmental whole-genome amplification to access microbial populations in contaminated sediments. *Applied Environmental Microbiology*, 72, 3291–330.

Amann, R.I, Ludwig, W. Schleifer.K.H. (1995). Phylogenetic identification and in situ detection of individual microbial cells without cultivation. *Microbiological Review*, 59, 143–169.

Angly,F.E. Felts,B, Breitbart,M, Salamon,P. et al. (2006). The marine viromes of four oceanic regions. *PLoS Biol*, 4, e368.

Beja,O. Koonin,E.V. Aravind,L. Taylor,L.T. et al. (2002). Comparative genomic analysis of archaeal genotypic variants in a single population and in two different oceanic provinces. *Applied Environmental Microbiology*, 68, 335–345.

Bertrand,H. Poly,F. Van,V.T. Lombard,N. Nalin,R. Vogel,T.M. Simonet,P. (2005). High molecular weight DNA recovery from soils prerequisite for biotechnological metagenomic library construction. *Journal of Microbiological Methods*, 62, 1–11.

Bodelier,P.L.E. et al. (2009). Whole community genome amplification (WCGA) leads to compositional bias in methane-oxidizing communities as assessed by pmoA-based microarray analyses and QPCR. *Environmental Microbiology*, Rep. 1, 434–441.

Burgmann,H. Pesaro,M. Widmer,F. Zeyer,J. (2001). A strategy for optimizing quality and quantity of DNA extracted from soil. *Journal of Microbiological Methods*, 45, 7–20.

Desai,C. Datta,M. (2006). Extraction of inhibitor-free metagenomic DNA from polluted sediments, compatible with molecular diversity analysis using adsorption and ion-exchange treatments. *Bioresource Technology*, 98, 761–768.

Field,D. Garrity,G. Gray,T. Morrison,N. Selengut,J. et al. (2008). The minimum information about a genome sequence (migs) specification. *Nature Biotech,* 26, 541–547.

Fiers,W. Contreras,R. Duerinck,F. Haegeman,G. Iserentant,D. et al. (1976). Complete nucleotide sequence of bacteriophage ms2 RNA: primary and secondary structure of the replicase gene. *Nature*, 260, 500–507.

Fleischmann,R.D. Adams,M.D. White,O. Clayton,R.A. Kirkness,E.F. et al. (1995). Whole-genome random sequencing and assembly of haemophilus influenzae rd. *Science*, 269, 496–512.

Garrity,G.M. (2005). Bergey's manual of systematic bacteriology. *Springer*.

Ginolhac,A. Jarrin,C. Gillet,B. Robe,P. et al. (2004). Phylogenetic analysis of polyketide synthase I domains from soil metagenomic libraries allows selection of promising clones. *Applied Environmental Microbiology*, 70, 5522–5527.

Guan,C. Ju,J. Borlee,B.R. Williamson,L.L. Shen,B. Raffa,K.F. Handelsman,J (2007). Signal mimics derived from a metagenomic analysis of the gypsy moth gut microbiota. *Applied Environmental Microbiology*, 73, 3669-3676.

Handelsman,J. Rondon,M.R. Brady,S.F. Clardy,J. Goodman,R.M. (1998). Molecular biological access to the chemistry of unknown soil microbes: a new frontier for natural products. *Chem Biol*, 5, R245–R249.

Hinoue,M. Fukuda,K. Wan,Y. Yamauchi,K. Ogawa,H. Taniguchi,H. (2004). An effective method for extracting DNA from contaminated soil due to industrial waste. *Journal of University of Occupational and Environmental Health* (Japan), 26, 13–21

Holben,W.E. (1994). Isolation and purification of bacterial DNA from soil. In: Weawer, R.W. (Ed.), Methods of Soil Analysis, Part 2. Soil Science Society of America, Inc, pp. 727–751.

Kauffmann,I.M. Schmitt,J. Schmid,R.D. (2004). DNA isolation from soil samples for cloning in different hosts. *Applied Microbiology and Biotechnology*, 64, 665–670.

Kottmann,R. Gray,T. Murphy,S. Kagan,L. Kravitz,S. et al. (2008). A standard MIGS/MIMS compliant xml schema: toward the development of the genomic contextual data markup language (gcdml). *OMICS*, 12, 115–121.

Morimoto,S. Fujii,T. (2009). A new approach to retrieve full lengths of functional genes from soil by PCR-DGGE and metagenome walking. *Applied Microbiology and Biotechnology*, 83, 389-396.

Pace,N.R. (1997). A molecular view of microbial diversity and the biosphere. *Science*, 276, 734–740.

Pace,N.R. Stahl,D.A. Lane,D.J. Olsen,G.J. (1985). Analyzing natural microbial populations by rRNA sequences. *ASM News*, 51, 4-12.

Picard,Ch. Ponsonnet,C. Paget,E. Nesme,X. Simonet,P. (1992). Detection and enumeration of bacteria in soil by direct DNA extraction and polymerase chain reaction. *Applied Environmental Microbiology*, 58, 2717–2722.

Pruitt,K.D. Tatusova,T. Maglott,D.R. (2005). NCBI Reference Sequence (RefSeq): A curated non-redundant sequence database of genomes, transcripts and proteins. *Nucleic Acids Research*, 33, D501–D504.

Quaiser,A. Ochsenreiter,T. Lanz,C. Schuster,S.C. et al. (2003). Acidobacteria form a coherent but highly diverse group within the bacterial domain: Evidence from environmental genomics. *Molecular Microbiology*, 50, 563–575.

Rajendhran,J. Gunasekaran,P. (2008). Strategies for accessing soil metagenome for desired applications. *Biotechnology Advances*, 26, 576–590.

Rappe´, M.S. Giovannoni,S.J. (2003). The uncultured microbial majority. *Annual Review of Microbiology*, 57, 369–394.

Riaz,K. Elmerich,C. Moreira,D. Raffoux,A. Dessaux,Y. Faure,D. (2008). A metagenomic analysis of soil bacteria extends the diversity of quorum-quenching lactonases. *Environmental Microbiology*, 10, 560-570.

Rondon,M.R. August,P.R. Bettermann,A.D. Brady,S.F. Grossman,T.H. et al.(2000). Cloning the soil metagenome: a strategy for accessing the genetic and functional diversity of uncultured microorganisms. *Applied Environmental Microbiology*, 66, 2541–2547.

Sanger,F. Coulson,A.R. Friedmann,T. Air,G.M. Barrell,B.G. et al. (1978). The nucleotide sequence of bacteriophage phix174. *Journal of Molecular Biology*, 125, 225–246.

Schleper,C. DeLong,E.F. Preston,C.M. Feldman,R.A. et al. (1998). Genomic analysis reveals chromosomal variation in natural populations of the uncultured psychrophilic archaeon Cenarchaeum symbiosum. *Journal of Bacteriology*, 180, 5003–5009.

Schmidt,T.M. DeLong,E.F. Pace,N.R. (1991). Analysis of a marine picoplankton community by 16S rRNA gene cloning and sequencing. *Journal of Bacteriology*, 173, 4371-4378.

Smalla,K. Creswell,N. Mendonca-Hagler,L.C. Wolters,A. VanElsas,J.D. (1993). Rapid DNA extraction protocol from soil for polymerase chain reaction-mediated amplification. *Journal of Applied Bacteriology* 74, 78–85.

Spencer, D.H. Kas, A. Smith,E.E. Raymond,C.K. Sims,E.H. Hastings,M. Burns,J.L. Kaul,R. and Olson,M.V. (2003). Whole-Genome Sequence Variation among Multiple Isolates of Pseudomonas aeruginosa. *Journal of Bacteriology*, 185(4), 1316-1325.

Tang,K. Kobayashi,R.S. Champreda,V. Eurwilaichitr,L. Tanapongpipat,S. (2008). Isolation and characterization of a novel thermostable neopullulanase-like enzyme from a hot spring in Thailand. *Bioscience Biotechnology and Biochemistry*, 72,1448-1456.

Tsai,Y.L. Olson,B. (1992). Rapid method for separation of bacterial DNA from humic substances in sediments for polymerase chain reaction. *Applied and Environmental Microbiology* 58, 2292–2295.

Uchiyama,T. Abe,T. Ikemura,T. Watanabe,K. (2005). Substrate-induced gene-expression screening of environmental metagenome libraries for isolation of catabolic genes. *Nature Biotechnology*, 23,88-93.

Venter,J.C. Remington,K. Heidelberg,J.F. Halpern,A.L. Rusch,D. Eisen,J.A. Wu,D. Paulsen, I. Nelson,K.E. Nelson,W. Fouts,D.E. Levy,S. Knap,A.H. Lomas,M.W. Nealson,K. White, O. Peterson,J. Hoffman,J. Parsons,R. Baden-Tillson,H. Pfannkoch,C. Rogers,Y.H. Smith,H.O. (2004). Environmental genome shotgun sequencing of the Sargasso Sea. *Science*, 304, 66–74.

Warnecke,F. Luginbühl,P. Ivanova,N. Ghassemian,M. Richardson,T.H. Stege,J.T. Cayouette, M. McHardy,A.C. Djordjevic,G. Aboushadi,N. Sorek,R. Tringe,S.G. Podar,M. Martin,H.G. Kunin,V. Dalevi,D. Madejska,J. Kirton,E. Platt,D. Szeto,E. Salamov,A. Barry,K. Mikhailova,N. Kyrpides,N.C. Matson,E.G. Ottesen,E.A. Zhang,X. Hernández,M. Murillo,C. Acosta,L.G. et al. (2007). Metagenomic and functional analysis of hindgut microbiota of a woodfeeding higher termite. *Nature*, 450,560-565.

Xu,M. Xiao,X. Wang,F. (2008). Isolation and characterization of alkane hydroxylases from a metagenomic library of Pacific deep-sea sediment. *Extremophiles*, 12,255-262.

Yooseph,S. Sutton,G. Rusch,D.B. Halpern,A.L. et al. (2007). The Sorcerer II Global Ocean Sampling Expedition: Expanding the universe of protein families. *PLoS Biol*, 5, e16.

Yun,J. et al. (2004). Characterization of a novel amylolytic enzyme encoded by a gene from a solid-derived metagenomic library. *Applied. Environmental. Microbiology*, 70, 7229–7235

Zhou,J. Bruns, M.A. Tiedje,J.M. (1996). DNA recovery from soils of diverse composition. *Applied Environmental Microbiology*, 62, 316–322.

In: Molecular Biology of Bacteria
Editors: R. Kumar Gaur and H. K. Gautam
ISBN: 978-1-62618-189-2
© 2013 Nova Science Publishers, Inc.

Chapter 3

Molecular Biology of Nitrile Metabolizing Enzymes

Komal Saxena[1], Ashish Deep Gupta[1], Girijesh Kumar Patel[2] and Vikash Babu[2*]
[1]Institute of Biomedical Education and Research,
Mangalayatan University, Aligarh – Uttar Pradesh, India
[2]Department of Biotechnology, Graphic Era University,
Clement Town, Dehradun, Uttarakhand, India

Abstract

Biotransformations involve the use of enzymes and microbes to carry out biochemical reactions. Nitrile metabolizing enzymes are such biocatalysts that have been emerged as an efficient tool in bioremediation of nitriles and in the production of industrially important organic acids and amides with various advantages such as mild reaction conditions, high catalytic efficiencies with reduced waste, ease of immobilization for the product recovery and recycle, tolerance to organic solvents and eco-friendly. In this chapter, genomic organization, expression and effect of mutations on the production of nitrile metabolizing enzymes have been explored.

Keywords: Nitrilase, NHase, amidase, nitrilases gene expression

1. Introduction

Biotransformation of nitriles takes place through two enzymatic pathways in bacteria. In the first pathway, nitrilase catalyse the direct transfer of nitriles to yield acid of the corresponding nitrile together with release of ammonia as the secondary product while in the second pathway, NHase is responsible for the direct clevage of nitriles to amides. Thereafter, this amide transforms to industrially important carboxylic acids by amidases (Figure 1). In general, bacteria exhibit either nitrilase catalyzed pathway or nitrile hydratase/amidase pathway. In some organisms for example *Rhodococcus sp.* both types of the nitrile degrading pathways are expressed depending on the choice of inducer for the expression of the respective pathway because these enzymes are generally inducible in nature. Furthermore, some bacteria contain both constitutive and/or inducible nitrile metabolizing enzymes (Brandão et al., 2003; D'Abusco et al., 2005; Kaplan et al., 2011; Kobayashi et al., 1992a; Kolkenbrock, et al., 2006; Pandey et al., 2009; Robertson, et al., 2004; Sosedov et al., 2010). Moreover, both NHases and Nitrilases are also involved in the biosynthesis of the plant hormones such as Indole acetic acid (IAA) and also present in plant-associated symbiotic microorganisms (Chauhan et al., 2003; Kato et al., 2005; Kiziak et al., 2009; Kobayashi et al., 1995).

Nitrilases (EC 3.5.5.1) are ubiquitous hydrolytic enzymes, belonging to a subfamily of the carbon-nitrogen hydrolase (C-N hydrolases) superfamily. These enzymes are found widely in nature amongst prokaryotes such as proteobacteria, actionobacteria, cyanobacteria in diverse range of habitats like soil, costal marine sediments, deep sea sediments and even in geothermal environments.

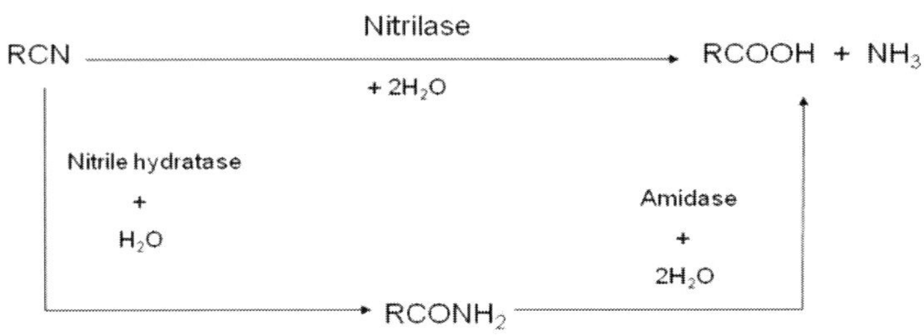

Figure 1. Biotransformation of nitriles by nitrile metabolizing enzymes.

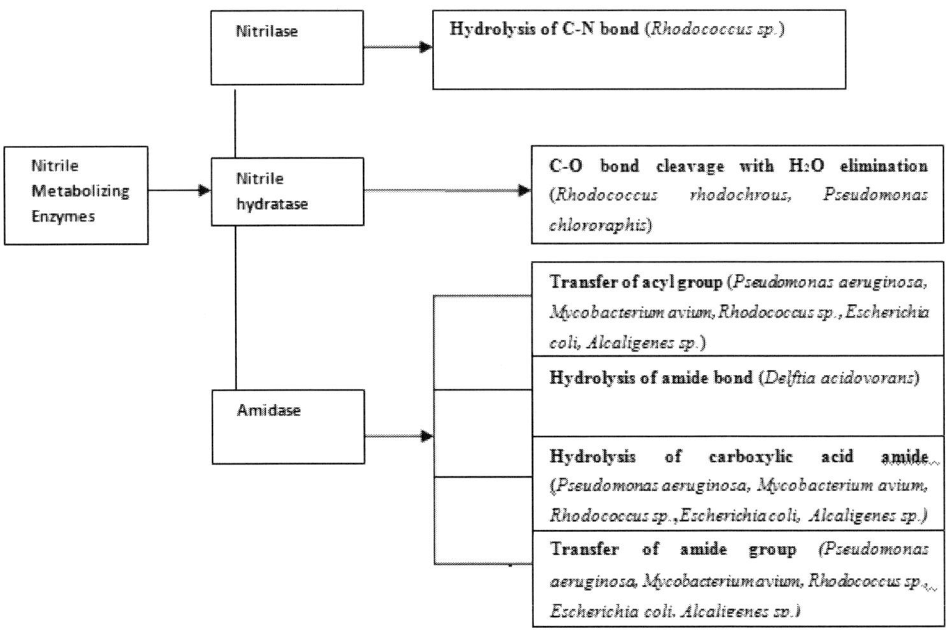

Figure 2. Reaction pathways for metabolism of nitriles in Bacteria.

Another type of nitrile metabolizing enzyme, nitrile hydratase (EC4.2.1.84) has been explored as an important industrial catalyst for the synthesis of polymers and chemicals. It is preferred for waste water remediation for the removal of nitrile herbicides which are neurotoxic, potent carcinogenic, and mutagenic in nature (Brandão et al., 2003; D'Abusco et al., 2005; Foerstner et al., 2008; Kaplan et al., 2011; Kolkenbrock, et al., 2006; Pandey et al., 2009; Payne et al., 1997; Robertson, et al., 2004; Shi et al., 2004; Sosedov et al., 2010) due to the presence of cyano group. Therefore, these enzymes have attracted the attention in biocatalysis for organic chemical processing in the production of commercially important products due to the environmentally friendly bioconversion in mild reaction conditions, quantitative yields which due to their enantioselectivity and regioselectivity. The nitrilase, nitrile hydratase/amidase belonging to different bacterial genus/species have different rections for biocatalysis of various nitriles and amides. For example, nitrilase of *Alcaligenes faecalis* and, *Alcaligenes sp.* metabolizes nitriles by hydrolysis of nitrile group. Nitrile Hydratase of *Rhodococcus rhodochrous, Pseudomonas chlororaphis* metabolizes nitriles by C-O bond cleavage with H_2O elimination while amidase of *Pseudomonas aeruginosa, Mycobacterium avium, Rhodococcus sp., Escherichia coli, Alcaligenes sp.* and *Pseudomonas aeruginosa, Mycobacterium avium, Rhodococcus sp.,* metabolize amides by different pathways (Figure 2).

2. Versitality of Nitrile Metabolizing Enzymes

The versitality in the biochemical properties of different gene clusters of the nitrilase amongst different organisms have evolved over time due to construction of novel pathways for the metabolism by gene duplication, natural selection and neo functionalization (Foerstner et al., 2008; Kato et al., 2005; Podar et al., 2005; O'Mahony et al., 2005). The microbial nitrile metabolism could also be a consequence of ecological relationships between eukaryotes and prokaryotes. Henceforth, the genes of the nitrilase superfamily could have horizontally transferred from plants to bacteria and archaebacteria in course of this eukaryotic-prokaryotic ecological relationship. This is known as horizontal gene transfer (HGT) (Robertson et al., 2004).

- Effect of horizntal gene transfer-

- The organisms which have more than one nitrilase are taxonomically unrelated such as *Pseudomonas sp., Klebsiella pneumoniae* and *Burkholderia fungorum*. It infers that this is a result of ancient gene duplications or horizontal gene transfer (Podar et al., 2005) whereas, the organisms harbouring two nitrlases which are phylogenetically related is a consequence of gene duplication.
- It has also been observed that the nitrilases belonging to different species do not obey the actual taxonomy of their host organism. The nitrilases in species which contain more than one nitrilases such as in *Pseudomonas sp., Klebsiella pneumoniae sp., and Burkholderia fungorum* is a result of ancient gene duplications or HGT. The HGT/ internal chromosomal rearrangements is evident from the nitrilase cluster in *Photorhabdus sp. in which* DNA elements (transposases, IS elements) are present just downstream of the *Nit1C* gene clusters in *Synechocystis sp.* and *Photorhabdus*

sp. and the flanking region of *Nit1C* gene which does not harbour any conserved sequence showing homology/similarity to pathway of genes from any other organism's gene/operon (Podar et al., 2005).

- The occurrence of NHase in eukaryotes as a fusion protein is also a result of HGT from probacteria (Foerstner et al., 2008).

2.1. Structural Organization of Genes

Structural organization of genes for nitrile metabolizing enzymes have been explored in many micro-organsims such as nitrilase gene of *R. rhodochrous* J1. It contains an open reading frame (ORF) which starts with methionine and ends with TGA. (Kobayashi et al., 1992a).

The nucleotide sequence of the nitrilase gene of *Acidovorax facilis* 72W contain an ORF having a presumed ribosomal binding site of five nucleotides at the upstream of the respective start codon (Chauhan et al., 2003). In many *Rhizobium sp,* a large plasmid (90 to 300 Mda) encodes genes for nitrogen fixation and nodulation by the metabolism of nitrilases whereas, in some pathogenic bacterium like *Agrobacterium tumefaciens*, the Ti-plasmid encodes genes for the synthesis of indole acetic acid (IAA) (Kobayashi et al., 1995).

Structural gene organization of nitrile hydaratase is different from nitrilases because these enzymes have α and β subunits in equimolar amounts with a basic stoichiometry of α, β and M (Fe or Co).

These subunits along with metal ions play an important role in -CN hydration. In addition to their function in the active center, the metal ions may play a role in enhancing the folding or the stabilization of the subunit polypeptides of the enzyme (Banerjee et al., 2002).

The primary sequences are well conserved among all known NHases while there is no apparent homology between two subunits (Kobayashi et al., 1992b). The α-subunit and the β-subunit encoding genes are adjacent in the same operon though their sequence within the respective operon varies. The NHase gene cluster in enantioselective *Rhodococcus erythropolis AJ270* contain *nha3* gene which is present downstream of the *nha2* gene for expression of NHase β subunit which encodes the activator protein P44k of the NHase pathway (Komeda et al., 2006; Barandao et al., 2003,. O'Mahony et al., 2005). The Cobolt (Co) containing acetonitrile hydratase (ANHase) of *Rhodococcus jostii* RHA1 contains two structural genes *anhA* and *anhB* encoding ANHase and they are separated from each other by the *anhE* gene (Okamoto et al., 2010).

The sequences flanking NHase genes encode activators responsible for the transportation of iron or cobalt ions to the active site of the NHase for its heterologous gene expression (Stevens et al., 2003).

The open reading frames at upstream and downstream of NHase structural genes are found to encode amidase structural genes hence, the genes encoding amidase, NHase, and some other proteins are part of an operon (Jin et al., 2010; Matamá, et al., 2007; Wu et al., 1997). The amidase gene together with α-subunit and β-subunit gene of NHase, P47K and *OrfE* are sometimes found to be encoded by a single operon (Banerjee et al., 2002; Nishiyama et al., 1991). Moreover, in *Pseudomonas chlororaphis* B23 two ORFs of genes expression P47K and OrfE lie just downstream of the β-subunit gene of the NHase which encodes five proteins (amidase, subunits a and J1, P47K, and Orf E) by the genes cotranscribed by a single mRNA (Banerjee et al., 2002; Nishiyama et al., 1991). Furthermore, the NHase gene of *Pseudomonas chlororaphis* B23 reveals the presence of an amidase just upstream of α-subunit of the NHase homologous to amidase of *Rhodococcus sp.* N-774 (Nishiyama et al., 1991).

The *amiEBCRS* amidase operon of *Pseudomonas aeruginosa* contain five genes namely-*amiB, amiS, amiE, amiC,* and *amiR* in which the regulatory *amiR* gene of the operon lies 2 kb downstream from the structural gene *amiE* (Wilson et al., 1995). The NHase and amidase expressing genes are also present in *Rhizobium sp.* (Kobayashi et al., 1995). The NHase pathway in the *Rhizobium sp.* is utilized for the production of IAA from tryptophan via indole- 3-acetamide. Such pathways do not occur in all the *Rhizobium sp.* for example, NHase pathway is present in a slow growing *Bradyrhizobium sp.* whereas, it is absent in fast-growing *Rhizobium sp.* (Kobayashi et al., 1995). These pathways are also present in some *Rhizobium leguminosarum*. For example, occurrence of amidase and NHase genes in *Rhizobium leguminosarum* MAFF 02-10231 and *Rhizobium. leguminosarum* MAFF 03-03021 (Kobayashi et al., 1995).

The ORF of some thermostable bacteria such as *Geobacillus pallidus* encodes an amidase of 348-amino acid residues found just at downstream of the α-subunit and the β-subunit of NHase (Makhongela et al., 2007). Although cyanobacterium rarely utilizes nitriles, for example cyanobacterium *Synechocystis* sp. PCC6803 is known to be one of the rare species that are able to utilize nitriles due an unexpected occurrence of nitrile metablizing gene in it (Heinemann et al., 2003).

3. Gene Expression

3.1. Induction

There are certain inducers which have been reported for the induction of many bacterial nitrilases (Komeda et al., 1996). The production of a recombinant nitrilase expressed in *Escherichia coli* JM109/pNLE in batch culture is induced by 0.2 mM isopropyl-*β*-D-thiogalactoside in medium (Liu et al., 2011). The 1.4 Kb region downstream from a nitrilase gene- *niL4* of an actinomycete *Rhodococcus rhodochrous* Jl encodes nitR gene whose product is a transcriptional positive regulator in *nitA* expression used in the induction of isovaleronitrile-dependent nitrilase synthesis using a *Rhodococcus-Escherichia coli* shuttle vector pK4. In alkaliphilic bacterium *Pseudomonas pseudoalcaligenes* CECT5344, the *nit1C* gene cluster is induced by cyanide in medium (Estepa et al., 2012). In the thermophilic bacterium *Bacillus pallidus* Dac521 meta-/para- substituted benzonitrile in medium induces nitrilase. Furthermore, medium containing isovaleronitrile is found to be optimal for the production of benzonitrilase by *Rhodococcus rhodochrous* J1 (Nagasawa et al., 1988). Similarly, *Alcaligenes faecalis* JM3 nitrilase is induced by addition of isovaleronitrile in medium (Mauger et al., 1990).

Similarly, bacteria harbor nitrile hydratase for the catabolism of nitriles which is inducible by amides, not by nitriles like isovaleronitrile (Kobayashi et al., 1995) i.e. *Rhodochrous sp.* J1has got two enzymes for the degradation of nitriles which are selectively induced by the use of urea and cyclohexanecarboxamide in the presence of cobalt ions. In this strain an interesting phenomenon of versatile nitrile metabolism exists where the metabolic pathway completely changes if the inducer is changed (Kobayashi et al., 1992a). In *Pseudomonas chlororaphis* B23 the NHase expression highly inducible by methacrylamide by virtue of *nhpR* gene regulator (Sakashita et al., 2008). In *R. erythropolis* JCM6823 and *R. erythropolis* JCM2892, NHase is induced by epsilon-caprolactam although, NHase in these strains is homologous to a constitutive NHase of *Rhodococcus sp*. N-774 that does not require an inducer (Duran et al., 1993).

The non consensus N-terminal regions of the XylS/AraC family regulators for NHase gene expression such as that of the *nhpR gene in P. chlororaphis* B23 is responsible for the recognition of various inducers as well as for the binding of activators for gene expression. The *nhpR* gene is oriented in opposite direction of the structural genes of aldoxime dehydratase, amidase, the α-subunit and β-subunits of NHase, NhpC, NhpS, and acyl-CoA synthetase (Sakashita et al., 2008). Whereas, strong induction of the NHase gene cluster by methacrylamide in *Pseudomonas chlororaphis* B23 could be needed for immediate catabolism of nitriles as they are very toxic in nature due to the presence of cyano group (Sakashita et al., 2008).

In *Rhizobium leguminosarum* MAFF 02-10231 and *Rhizobium leguminosarum* MAFF 03-03021, amidase and NHase genes are induced in the presence of crotonamide and isovaleronitrile respectively. Moreover, the NHase in both of them shows similarity to the NHase from *Pseudomonas chlororaphis* B23 in their four subunits which are identical to its NHase (Kobayashi et al., 1995).

3.2. Activation

Nitrilases do not require any metal co-factor or prosthetic group like Fe-type NHase or Co-type NHase (Banerjee et al., 2002). However, Mg^{2+} supplemented medium shows increased production of nitrilase and growth of biomass in *R. erythropolis* ZJB-0910 (Miller et al., 1982).

The Fe-type NHase prefers small aliphatic nitriles as substrates whereas the Co-type NHase prefers aromatic nitriles as substrates (Nojiria et al., 2000). Some NHase might require both iron and cobalt for its activity (Payne et al., 1997). NHase of *Agrobacterium tumefaciens* contained both cobalt and iron ions, rather than cobalt-type NHase or iron-type NHase (Kobayashi et al., 1995). *Rhodococcus* sp. N-774 and *Pseudomonas chlororaphis* B23 require ferric ions for their activity (Takashima et al., 2000). On the other hand, NHase of *B. smithii* SC-JOS-1 requires non coordinated manganese ions together with Cobalt ions for its activity. Although the role of non-coordinated metal ions is still not clear (Takashima et al., 2000).

An enantioselective NHase from *Rhodococcus erythropolis* AJ270 could be expressed efficiently in *E. coli* by the use of an activator protein P44k which is encoded as a separate ORF from NHase on a single replicon because, this activator protein could be involved in correct protein folding as it has been observed that the presence of iron as enhancer in absence of P44k does not lead to enhancement of enzymatic activity. However, P44k is also hypothesized to have a key role in Fe trafficking in NHase biogenesis to become a Fe type metallochaperon (Komeda et al., 2006). Among amidases, Co^{2+} is the activator of the L-stereoselective amidase - *laaA*$_{Bd}$ of *Brevundimanas diminuta* (Komeda et al., 2006) whereas, the sAmiD *N*-acetylmuramoyl-l-alanine amidase - zinc metalloenzyme binds single divalent zinc ion in its active site (Pennartz et al., 2009).

3.3. Enhancement

Many compounds or organic solvents have been reported for the enhancement of activity. The surface modification of an acrylic fibre was done by the use of nitrilase. Herein, the activity of the nitrilase is enhanced by a combinatorial effect of 1 M sorbitol and 4% N, N-dimethylacetamide if added to the medium (Matamá et al., 2007). Moreover, the addition of water miscible organic solvents with immobilized nitrilase upto a critical concentration also results in enhancement in the rate of reaction because the affinity (K_m) of the nitrilase for its substrate depends on the property of aprotic solvent used which shows that how solvent control the enzymatic enantioselectivity by modulating the traffic of substrate at the active site of nitrilase (Kaul et al., 2008). However, enantioselective nitrilase of *E. coli* is inhibited at substrate concentrations >300 mM leading to a relatively lesser yield of (*R*)-(−)-mandelic acid. The problem could be solved so as to achieve enhanced biocatalysis by the use of water–organic solvent biphasic systems. The best of such biphasic system is one which contains toluene as the organic phase (Zhang et al., 2011).

The thiol group containing compounds such as dithiothreitol and 2-mercaptoethanol enhances the enzymatic activity upto its maximal in inducible nitrilase of *Alcaligenes faecalis* JM3. It has also been observed that, in *Alcaligenes faecalis* CCTCC M 208168 during bioconversion of (*R*,*S*)-mandelonitrile (*R*,*S*-MN) to (*R*)-(−)-mandelic acid (*R*-MA) by nitrilase via whole cells biocatalysis, the product- *R*-MA inhibits its own production and therefore requires *in situ* product removal (ISPR) by anion-exchange resin so to improve the *yield of the product.*

Enhancement of the NHase activity was done by the addition of n-butyric acid and $FeSO_4$ in enantioselective NHase from *Rhodococcus erythropolis* AJ270 (Komeda et al., 2006). Similarly, the catalysis of the Glu-Cys-Lys catalytic traid containing enzmes of the nitrilase superfamily such as RAPc8 amidase from *Geobacillus pallidus* is enhancd upto 1.5 fold in the presence of Dithiothreitol (DTT) in the reaction mixture due to the presence of sulfhydryl groups on the catalytic site of the enzyme (Makhongela et al., 2007).

3.4. Regulation

Komeda et al.*(*1996), reported the presence of of an ORF corresponding to a 957 bp *nitR* gene encoding a positive transcriptional regulator for *nitA* gene expression in the downstream region of nitrilase gene *nit A* of 1.4 kb in *Rhodococcus rhodochrous* is responsible for the induction nitrilase gene expression. The transcriptional regulator gene *nitR* of *Rhodococcus rhodochrous* J1 gene is homologous to *araC* gene located just downstream of *nitA* in *E. coli* whereas, in *Bacillus* OxB-1 sp. nitrilase gene is connected to gene for phenylacetaldoxime dehydratase expression (Kiziaik et al., 2005). On the other hand, the nitrilase gene in *Pseudomonas fluorescens* is connected to the gene of mandelate pathway which is a mandelonitrile metabolizing enzyme producing gene that converts mandelonitrile to mandelic acid (Kiziaik et al., 2005). A ''PnitA-NitR'' system has been developed for the regulation of nitrilase gene expression in *Streptomycetes sp.,* which is based on the inducible pathway (inducible with e-caprolactam) of nitrilase of *R. rhodochrous* J1 (Herai et al., 2004). The 1.4 Kb region from the downstream of a nitrilase gene *nitA* of an actinomycete *Rhodococcus rhodochrous* Jl encodes *nitR* gene whose product is a transcriptional positive regulator in *nitA* expression is used in the induction of isovaleronitrile-dependent nitrilase synthesis using a *Rhodococcus-Escherichia coli* shuttle vector pK4 because, this region has been found to be consisting of an open reading frame of nitrilase *nitR* gene of 957 bp. The deletion of central and 3'- terminal portion of *nitR* gene leads to elimination of nitrilase activity and the sequence similarity search also reveals the amino acid sequence of *nitR* similar to positive regulator family sequences such as including XylS of *Pseudomonas putida* and AraC of *E. coli* (Komeda et al., 1996).

The Xyls/AraC family regulators are positive regulators of metabolism of various carbon sources and harbor sequence similarity in their C- terminus by having a characterstic helix-turn-helix motif for DNA-binding (Komeda et al., 1996). The 5' upsteam region of *oxdA*, contains an ORF expressing the positive transcriptional regulator of the NHase gene expression *nhpR* belonging to XylS/AraC family like the *nitR* gene expression for the metabolism of nitrile in *Rhodococcus rhodochrous* J1 (Komeda et al., 1996; Sakashita et al., 2008). Furthermore, in *Pseudomonas chlororaphis* B23 the NHase expression is highly inducible by methacrylamide. Its gene cluster comprising of seven genes namely *oxdA*, *amiA*, *nhpA*, *nhpB*, *nhpC*, *nhpS*, and *acsA* where, the first five genes *oxdA*, *amiA*, *nhpA*, *nhpB*, and *nhpC* are cotranscribed and the later occurring genes *nhpS* and *acsA* are also cotranscribed but, the high level of gene induction in presence of methacrylamide exists due to the presence of *nhpR* gene (in opposite orientation to the above described genes) located upstream of *oxdA* gene which is the positive regulator of NHase gene cluster encoding a transcriptional activator responsible for NHase gene cluster expression control regulation belonging to XylS/AraC transcriptional gene regularor family including *NitR, XylS1*and *CadR* genes expression products which are the regulators of nitrile metabolism in *Rhodococcus rhodochrous* J1(positive regulator), *Pseudomonas putida* (regulator for alkylbenzoate metabolism) and in *Bradyrhizobium* sp. HW13 (the *cad* gene regulator) respectively (Komeda et al., 1996; Sakashita et al., 2008). *Rhodococcus pyridinovorans* contain gene encoding NHase which is homologous to NHase of *Rhodococcus rhodochrous J1* with only difference in the gene sequence of the two NHase lies in the regulatory sequence although, this minute difference in the structue of the the two NHase leads to a decrease of NHase activity in *R. pyridinovorans* because it exibits only 20% of the specific activity of the H-NHase of *Rhodococcus rhodochrous J1*. The activity of both of the genes is strongly controlled by Co^{2+} and amides via regulatory gene product located within the operon of NHase gene expression (Precigou *et al.*, 2004). Similarly, NHase of *Rhodococcus pyridinovorans* which rarely show NHase activity is homologus to NHase of *Rhodococcus rhodochrous* J1 with only difference if any in the regulatory region of the enzyme. The operons for NHase production are under control of Co^{2+} and amide compounds influencing regulatory genes encoded by the operon for the NHase metabolism (Precigou et al., 2004). Furthermore, the activity of NHase from *Rhodococcus erythropolis* 9675 which transform herbicide 2,6-dichlorobenzonitrile to 2,6-dichlorobenzamide (BAM) could be down-regulated in the presence of high amount of nitrogen availability (NH_4Cl). Similarly, there are past reports which confer that NHase activity could be ceased by the presence of high amount of nitrogen availability ($(NH_4)_2SO_4$

like that in *Bacillus smithii* SC-J05-1 (Holtze *et al.*, 2006). Moreover, the σ^{54} is a regulatory factor is required for the expression of nitrogen metabolizing enzyme producing genes such as in nitrogen fixation. In a similar fashion, a consensus σ^{54}- dependent promoter sequences −24(TGGCACG)/−12(TTGCA) could be responsible for NHase gene expression. Since, aldoxime dehydratase, amidase, and NHase are also involved in nitrile metabolism via aldoxime → nitrile → amide pathway to form corresponding acid and ammonia as end products. Furthermore, the transcription ratio of the promoter region upstream of *nhpA* and *oxdA* controls the level of the expression of NHase gene cluster since, the regulation of gene expression occurs at the post transcriptional level. The high level of gene expression of NHase could also stabilize the protein by protecting it from proteolytic degradation (Sakashita et al., 2008). Amongst the *amiEBCRS* operon for amidases, *amiR* gene is the positively regulator of the amidase pathway which is 2 kb downstream of the *amiE* gene and transcribed in the same direction. Another regulatory gene *amiC* also occurs but it negatively regulates amidase expression (Wilson et al.,1995). Moreover, the positive regulation of the amidase pathway is by virtue of transcription anti-termination where, the the gene regulator - AmiR may allow RNA polymerase to override a rho-independent transcription terminator between the *amiE* promoter and structural gene (Wilson et al., 1995, Zhou et al., 2009).

3.4.1. Inhibition

Many compounds have been reported for the inhibition of nitrile metabolizing enzymes production or activity. In *Aspergillus niger* K10 *p*-hydroxymercuribenzoate as well as heavy metal ions are inhibitory (Kaplan et al., 2006). Moreover, metal ions other than Mg^{2+} also showed inhibition of the nitrilase in *R. erythropolis* ZJB-0910. In thermophilic bacterium *Bacillus pallidus* Dac521 the ortho-substituted inducer – benonitrile inhibits the reaction due to steric hindrance instead of inducing the nitrilase gene expression. Furthermore, the nitrilase gene of *Rhodococcus rhodochrous* J1 could be inactivated together with loss of of a sulfhydryl group enzyme subunit by a 5,5'-dithiobis(2-nitrobenzoic acid) modification (Kobayashi et al., 1992a). Kaul et al.(2008) observed that, the addition of water miscible organic solvents with immobilized nitrilase more than the critical concentration also leads to inhibition of the reaction instead of enhancing the rate of reaction. Moreover, the enantioselective nitrilase of *E. coli* is inhibited at substrate concentrations >300 mM leding to a relatively lesser yield of (*R*)-(−)-mandelic acid (Zhang et al., 2011).

The NHase is sensitive to sulfhydryl reagents due to the presence if Cys and Ser at the catalytic site of the enzyme whereas, oxidizing agents, chelating agents, and reducing reagents show only weak inhibition on the enzymatic activity (Feng et al., 2008). The enantioselective NHases from *Rhodococcus equi* A4 that prefers S-isomers of racemic 2-(2-, 4-methoxyphenyl) propionitrile, 2-(4-chlorophenyl) propionitrile and 2-(6-methoxynaphthyl) propionitrile (naproxennitrile) as substrates could be inhibited completely by 1 mM of Ag^+, H_2O_2, DTT and phenylhydrazine. It could also be inhibited to a large extent by iodoacetamide although; EDTA also leads to slight inhibition in the activity of the enzyme (Prepechalova et al., 2001). Furthermore, the mutations in NHase gene which could change the side-chain group of the enzyme can also result in total inactivation of the enzyme like those found in NHase in the *E. coli* BL21 (DE3)/pETNHM host expression system (Yu et al., 2006).

On the other hand, the Co^{2+} dependent $laaA_{Bd}$ amidase of *Brevundimanas diminuta* could be inactivated in the presence of EDTA due to chelation of the metal ions which are activators of this pathway like Co^{2+}, Ni^{2+} and Mn^{2+} (Komeda et al., 2006). Similarly, the AmiD which is a zinc metalloenzyme could be inactivated by EDTA (Kiziak et al., 2005; Pennartz et al., 2009). The amidase LamA from *O. anthropi* is not inhibited by serine proteases and cysteine protease because it is a metalloenzyme and hence could obviously be inhibited by metal ion chelators like EDTA although, the activity of the amidase could be restored by addition of the divalent metal ion Co^{2+}, Fe^{2+}, or Zn^{2+} (Sonke et al., 2005). Moreover, the catalysis of the Glu-Cys-Lys catalytic traid containing enzmes of the nitrilase superfamily such as in the *G. pallidus* RAPc8 amidase is inhibited by heavy metal ions Hg^{2+}, Fe^{2+}, and Co^{2+}. Furthermore, Fe^{3+}, Cu^{2+} and Zn^{2+} also inhibit the enzymatic activity. Inhibition and stability due to chelating agents, redox reagents, and specific covalent and non-covalent modification reagents is due to the presence of catalytically active cysteine thiol group. In some strains, ethylenediaminetetraacetic acid (EDTA) does not inhibit the enzymatic activity which shows that neither Ser residue nor coordinated metal ions in the catalytic/stabilizing sites play any role in the enzymatic activity (Makhongela et al., 2007).

4. Cloning and Overexpression

Escherichia coli is the best-known cloning vehicle for the expresion of nitrilase because the genes are sucessfully cloned and overexpressed in *E. coli.* The first nitrilase gene to be sucessfully cloned by Stalker and McBride in 1987 in *E.coli* was from a bromoxynil-degrading bacterium *Klebsiella pneumoniae* subsp. *Ozaenae*. Since then, many more genes for nitrile metabolism were cloned in *E. coli* e.g. the *nit*A gene from *R. rhodochrous* K22 in plasmid pNK21, *nitA* gene from *R. rhodochrous* in plasmid pNJ1 (Chauhan et al., 2003; Banerjee et al., 2002; Seffernick et al., 2009). pUC18 is used as a suitable vector having a 1.3-kb PuuI-NarI fragment inserted between SmaI and AccI sites for cloning of *pNJll* gene in *E.coli* under the control of the *lac* promoter. Hence, the nitrilase from *pNJll* gene in *E.coli* is only expreesed when IPTG is present in the culture medium. The nitrilase thus produced is in its active form whereas; NHase is produced in insoluble form in *E. coli* (Kobayashi et al., 1992b). In *E. coli* having pPCN4 gene for NHase expression is controled by *lac* promoter. The expression of H-NHase and L-NHase genes in *E. coli* is also under the control of a *lac* promoter in the medium containing $CoCl_2$ (Chauhan et al., 2003; Banerjee et al., 2002; Seffernick et al., 2009). The NHase of *Pseudomonas chlororaphis* B23, when cloned in *Escherichia coli* by the DNA-probing method together with the NHase gene of *Rhodococcus sp.* strain N-774 as the hybridization probe then, a sequence similar to the amidase of *Rhodococcus sp.* strain N-774 is found to be encoded by the region just upstream of the α subunit of NHase gene sequence. Furthermore, P47K and OrfE proteins are expressed by the region just downstream of the corresponding β subunit of NHase gene sequence. Here, the five gene expression products namely -amidase, α and β subunits, P47K, and OrfE are cotranscribed by a single mRNA. The plasmid pPCN4 for cloning is under the control of the lac promoter and isopropyl-β-D-thiogalactopyranoside enhances the activity of the NHase expression in *E. coli* (Nishiyama et al., 1991). A novel recombinant strain, *E. coli* BL21 (DE3)/pETNHM, has overexpression of NHase in the host expression system (Yu et al., 2006). The nitrilase gene in *Acidovorax facilis* 72W could be over expressed about 15-fold higher in *E. coli* under control of the T7 promoter in a pET vector so as to obtain high yields of 4-cyanopentanoic acid by 100% conversion of substrate together with no amide formation and negligible amount of 2-methylglutaric acid formation as the byproduct of metabolism (Chauhan et al., 2003; Banerjee et al., 2002; Seffernick et al., 2009).

Some amidases such as the Penicillin amidase (PA) could also be cloned and ovrexpressed in *E. coli* such as *E. coli* ATCC 11105 (Ignatova et al., 2000). The *laaA*$_{Bd}$ of *Brevundimanas diminuta* which is encoded by an open reading frame encoding 491 amino acid residue polypeptide could be overexpressed by modifying the *laaA*$_{Bd}$ sequence upstream from the start codon (Komeda et al., 2006). Thermostable amidase of thermophilic bacterium like *Geobacillus pallidus, Pseudonocardia thermophila* JCM 3095 and broad spectrum amidase from *B. stearothermophilus* BR388 could also be cloned and over expressed in *E. coli* (Makhongela et al., 2007; Banerjee et al., 2002). Research for the direct evolution and rational designing of the NHase are in focus so as to obtain thermostable, substrate and product tolerant strains by the use of recombinant *E. coli* XL1-Blue (pUC18-NHBAM). Thereby, the NHase genes was cloned reversibly in the direction of the origional *Nocardia sp.* by homologous recombination to double the strength of expression of NHase gene as well as for the destruction of the amidase gene in order to prevent the conversion of the product to some byproducts such acrylic acid (Shi et al., 2004). *Moreover, the* pyrazinamidase/nicotinamidase (PncA) of *Mycobacterium smegmatis* is also cloned and expressed in *E. coli (*Heinemann et al., 2003*)*. Sometimes *E. coli* fail to produce active NHases. One of such evidence is the formation of insoluable aggregates of the NHase genes expression products from *Rhodococcus* sp. N-774 and *Rhodococcus rhodochrous* (Banerjee et al., 2002). If the subunits K and L of the NHase are expressed in *E. coli* host vector system without the coexpression of NHase activator protein then the NHase subunits form inactive inclusion bodies (Nojiria et al., 2000). NHase and amidase genes of *Rhodococcus* sp. N-774 could be expressed in *R. rhodochrous sp.* For example, when NHase and amidase genes of *Rhodococcus* sp. N-774 were expressed in *R. rhodochrous* ATCC 12674, it produced active form of NHase and amidase (Banerjee et al., 2002). *Bacillus megaterium is known to be efficient in cloning and expression of amidase. This bacterium is advantageous than E.coli for the expression of secretory proteins such as Penicillin G* due to plasmid stability and a low intrinsic protease activity in it. Although, *E. coli* has a high capacity for protein export as well. Therefore, *Bacillus megaterium* is preferred than *E.coli* for the production of penicillin G amidase (PGA) due to higher synthesis rate than that of *E. coli. Bacillus megaterium secretes* PGA directly to the medium which reduces the cost of downstream processing (Yang et al., 2006).

5. Effect of Mutations/Alterations on Gene Expression

5.1. Effect of Mutations in C-Terminus

The desirable alterations in the structure of the nitrilase by vitrue of mutant variants can alter the substrate specificity and enantioselectivity because it has been observed that the C-terminus of the nitrilase is crucial in deciding its enentioselectivity (Kiziak et al 2009; Sosedov et al 2009; Kiziak et al., 2005). The C-terminus alterations significantly affects the activity and stability of the nitrilases togather with the loss of the preferred binding of one of the nitrile enantiomers sometime resulting in altered enantioselectivity. (Kiziak et al 2009; Sosedov et al 2009; Kiziak et al., 2005). The deletion in the C-terminus residues of the nitrilase dimnishes the activity of the enzyme. Larger the deletions consequently greater decrease in the activity of the nitrilase is observed. In the C-termini the amino acids residues 302 and 319 sequence are important for the enzymatic activity though they do not take part in the presumed catalytical triad of the enzyme (Kiziak et al., 2005). In *Pseudomonas fluorescens,* the Nit(del-C-47), Nit(del-C-55), Nit(del-C-60) and Nit(del-C-67) mutants which results in the structural change in the enzyme show increased activity of the enzyme whereas, the deletion mutants Nit(del-C-47)–Nit(del-C-75) changes the enantioselectivities (Kiziak et al., 2005).

5.2. Effect of Amino Acids Having Side Chains

In *Pseudomonas fluorescens,* the His296 mutant reveals that not only amino acid sequence at the C-terminal is responsible for various physiochemical properties of the nitrilase but the side chains of some amino acids in nitrilase structure also govern the catalytical properties because a basic residues at position 296 of nitrilase affects the enzymatic activity more than a neutral amino acid residues at the same position. Moreover, an increase in the size of the side chain of the amino acid in *R. rhodochrous* ATCC 33278 at position Tyr142 caused the decrease in the k_{cat} values for aromatic nitriles and the alteration of aromatic amino acid residues at position Tyr142 with positively charged amino acids leads to a total loss of activity. Similarly, the presence of non-polar aliphatic amino acids like Ala, Val, Leu at position Tyr142 results in selectivity of nitrilase for aromatic nitriles only due to the same reason (Yeom et al., 2008).

5.3. Effect of Mutations on Activity and Enantioselectivity

Sometimes homologous nitrilases in different organisms such as *Pseudomonas fluorescens* EBC191 and *Alcaligenes faecalis* ATCC 8750 show variability in their choice for same substrate. Hence, the same enzyme from *Pseudomonas fluorescens* EBC191 converts (*R*,*S*)-mandelonitrile with low enantioselectivity to (*R*)-mandelic acid and (*S*)-mandeloamide as side product whereas, nitrilase from *Alcaligenes faecalis* ATCC 8750 hydrolyses (*R*,*S*)-mandelonitrile to (*R*)-mandelic acid and no side product is formed (Kiziak et al., 2009). An Ala at position 165 in the NitP enzyme is responsible for a distinction in the causative mechanism of action of homologous and chimeric enentioselective nitrilases in *Pseudomonas fluorescens* EBC191 and *Alcaligenes faecalis* ATCC 8750 (Kiziak et al., 2009). The nitrilase of the former shows lower enantioselectivity between (*R*)-mandelic acid and (*S*)-mandeloamide as substrates whereas, in the later species the homologous nitrilase shows higher enantioseletivity for (*R*)-mandelic acid as substrate although, a chimeric enzyme constructed from both of them resembles *A. faecalis* ATCC 8750 in its turnover characteristics for (*R*,*S*)-mandelonitrile and (*R*,*S*)-2-phenylpropionitrile (2-PPN) and is more enantioselective for (*R*)-mandelic acid formation than that of *A. faecalis* (Kiziak et al., 2009).

There is difference in the active and inactive states of the NHase. The active state of NHase consists of αCys109, αCys109 and αCys1114 together with two amide nitrogen atoms coordinating to ferric iron in the Fe-type NHase at αSer113 and αCys114 whereas, in the inactive state two Cys residues are coordinated equatorial to iron so as to have post translational modifications forming Cys-sulfenic and Cys-sulfinic acids (Song et al., 2007). If *NhpR* gene belonging to regulators in *Pseudomonas chlororaphis* B23 is mutated then the activation of the nitrile gene cluster is lost. Hence, the pathway is no more induced by methacrylamide as inducer because nonconserved N-terminal region is thought to be responsible for binding to an activator as described previously in this chapter (Sakashita et al., 2008). The gene expression of NHase in *Rhodococcus erythropolis* DSM 43066^T is unique amongst all the *Rhodococcus erythropolis* species because a Pro residue at position 71 in the β subunit in this strain next to two highly conserved amino acids residues $Tyr72^B$ and $Tyr73^B$ makes this enzyme nonfunctional due to alterations in the structure of the enzyme (Brandao et al., 2003).

5.4. Effect of Mutations in the Conserved Regions on Enzymatic Activity

In nitrilase from *Alcaligenes faecalis* JM3 and *Klebsiella ozaenae* having only Cys-163 the most conserved amongst both, this residue is responsible for the catalytic activity of nitrilase in *Alcaligenes faecalis* JM3 because mutation leading to substitution of Cys163Ala leads to complete loss of the enzymatic activity (Kobayashi et al., 1993). Similarly, in *Pseudomonas fluorescens* EBC191, the Cys at the same amino acid position in the nitrilase sequence is responsible for its catalytic activity (Sosedov et al., 2010). Any mutation not only in the catalytic site but also near to it also affects the activity of the enzyme since these residues forms the bottom of the substrate-binding pocket of the enzyme. In *A. faecalis* ATCC 8750, one of such example is the mutations in the conserved sequence Cys-(Trp)-Glu-His being Glu166Ala, His167Ala results in 0.1% - 0.3% activity than that of the wild type enzyme for 2-PPN and mandelonitrile as substrates for mandelic acid formation. It has been also observed that certain mutations also lead to inversion of the enzyme activity because mutation NitA(Trp164Ala) in enantioselective nitrilase from *A. faecalis* ATCC 8750 show preferential formation of (*S*)-2-PPA from (*R,S*)-2-PPN which reveal the inversion of enzymatic activity than the wild-type enzyme NitA. This confers that a single amino acid change in the enzyme can decide the enantioselectivity of the enzyme because change in the amino acid can lead to stearic hinderance and charge-related interactions at the active site of the enzyme. The aromatic ring substitution of Phe at the active site identical to benzene ring of 2-PPN showed the highest degree of enantioselectivity for (*R*)-PPA formation suggesting that an aromatic amino acid residue exists at the cataytic site of the enzyme (Kiziak et al., 2009).

5.5. Some Mutations Do Not Alter the Specificity

Some mutations forming non polar aliphatic amino acid residues do not alter the specificity of nitriles. In nitrilase such as that of *Rhodococcus rhodochrous,* which hydrolyses both aliphatic and aromatic nitriles, some mutations in non polar aliphatic amino acid residues do not alter the specificity of the substrate.

In the strain of *Rhodococcus sp.,* mutation containing Ala, Val, Leu at amino acid residue 142 instead of Tyr does not change the specificity of the nitrilase for the substrate whereas, the mutations containing charged amino acids like Aspartate, Glutamate, Arginine and Aspargine at the same site leads to total inhibition of the enzymatic activity of the enzyme for nitrile metabolism due to disruption of hydrophobic interactions of the enzyme-substrate complexes. This infers that aromatic residue at position 142 is essential for the activity of the nitrlase in this bacterial strain (Yeom et al., 2008).

6. Consensus Sequence and Motifs Responsible for Catalysis

Nitrilases have characterstictic α-β-β-α sandwich fold and conseneseus motifs of the C-N hydrolase superfamily to which they belong. The nitrilase superfamily consists of conserved Glu-Lys-Cys catalytic traid where, the conserved Cys residue of the C-N hydrolase family acts a nucleophile for the attack of the substrate at the catalytic site of the enzyme (Seffernick et al., 2009). The consensus residues at the catalytic domain in some organisms contain Glu120, Lys278, and Cys329 having glutamate to be least conserved among them. Although, at Thr/Ser at position 283 the consensus sequence of nitrilases exists. In archaeal species Phe occurs at the same position in nitrilases (Seffernick et al., 2009). The nitrilase from *Akalgenes faecalis* JM3 and its homologue from *Klebsiella ozaenae* contain only Cys-163 which is the most conserved amino acid residue among both (Kobayashi et al., 1993; et al., Kobayashi et al., 1992a). The nitrilase from *Acidovorax facilis* 72W contain a conserved sequence of 50 amino acid residues in the polypeptide of the enzyme which play a vital role. This conserved amino acid sequence works together with the conserved signature domains for example 16 amino acid region near the amino terminus (40–55 72W numbering), and the catalytic region (160– 173 72W numbering) containing the essential Cys residue. These conserved sequences are also found in other nitrilases homologous to this nitrilase (Chauhan et al., 2003). The nitrilase from *Pseudomonas* sp. harbors a consensus motif D-(P/S)-X-G-H-Y between amino acid residues 273–298 (Kiziak et al., 2005).

NHase contains a hexapeptide residue core GDVSPN which lies in the middle of a larger conserved sequence, namely NXGDVSPNXXG$^{P}/_{X}$XC. The GDVSPN motif is also a consensus motif among various neutral ceramidase (CDases) which plays a key role in the catalytic activity of this enzyme governed by serine at the catalytic site (Galadari et al., 2006). Many phage- encoded peptidoglycans also share conservative amidases domain with a wide range of bacterial autolysins such as N-acetylmuramoyl-Lalanine amidases, bacterial glutathionylspermidine amidases, eukaryotic glutathionylspermidine amidases, g-D-glutamyl-L-diamino acid endopeptidase and NLP/P60 family proteins. Moreover, the conserved cysteine and histidine residues are found amongst all of them which is essential for the catalytic activity of the enzyme for the hydrolysis of g-glutamyl as substrates. Likewise, in *Vibro anguillarum* which is the causative agent of fish disease vibrosis contain amidase gene *amiB* encoding peptidoglycan hydrolase N-acetylmuramoyl-L-alanine amidase for the cleavage of peptidoglycan (Hong et al., 2007; Komeda et al., 2006). In contrast to Glu-Lys-Cys catalytic traid in nitrilase (Seffernick et al., 2009) the amidase consensus signature residues amongst prokaryotes and eukaryotes have GGSSGG or GGSS(S/G)GS motifs consisting of Gyl, Ser and Ala. In aliphatic amidases a conserved sequence motif Cys-Tyr-Glu or Cys-(Glu)-Glu-Gly exists at the catalytic site of the enzyme (Hashimoto et al., 1992). In contrast, a novel class of amidase called ceramidase; EC3.5.1.23 (CDase) does not contain the fatty amidase signature sequence $GSSG(X)_{14}GXD(X)_3$ $S(X)_3$ PX. Furthermore, a slighty different motif exists in some of the bacterial amidase i.e. the GASSSG motif in *D, acidovarons and V. paradoxus.* In contrast, formamidase from *Bacillus cereus* CECT 5050T (BceAmiF) contains the Glu140 and Tyr191 residues which are responsible for the enzymatic activity of the AmiF superfamily although, yet another glutamate residue is found amongst various members of the nitrilase superfamily involved in formation of the C-E-K triad i.e. a C-E-E-K tetrad in their conserved region. Furthermore, it has also been observed that amidases of some organisms such like *Pseudomonas aureginosa* and *O. anthropi* do not have any conserved motifs (Hongpattarakere et al., 2005; Ignatova et al., 2003).

6.1. Effects of Post Translational Modifications

Some post translational modifications are conserved. For example, all nitrilases contain a cysteine residue in their catalytical center which catalyzes the hydrolysis of their substrates (Odaka et al., 2001). Various other post translational modifications are also found in the Fe-type NHase and Co-type NHase (Odaka et al., 2001).

Some posttranslational modifications in the NHase, nitrilase and amidase which are critical for their catalytic activity are described as follows:

a. The NHase of the bacteria have posttranslational modifications at their metallocentre in the catalytic α-subunit and β-subunit containing mononuclear non-heme iron (III) or non-corrin cobalt (III) (Odaka et al., 2001). The Cys- SO_2H of Cys2 is conserved in H-NHase of *Rhodococcus rhodochorous* J1. some conserved postranslational modifications contribute in forming consensus sequences in the motifs like - Cys1-Xxx-Leu-Cys2-Ser-Cys3, Xxx = Ser (Fe-type)/Thr (Co-type) consenses sequence present in the motif of all known NHases.There are other novel post translational modifications found in Fe-type NHase of *Rhodococcus* sp. N771 having unusual structure of the iron center in nitrosylated state in which αCys112 and αCys114 are post-translationally modified to cysteine-sulfinic acid (Cys-SO2H) and –sulfenic acid (Cys-SOH), respectively. Due to this post-translational modification at the cysteine residues, this strain shows a unique photoreactivity reaction though the phenomenom of photoreactivity which is found commonly in Fe-type NHase of *Rhodococcus sp.* This shows that the post translational modifications play a key role in catalysis such that the enzyme is inactive if a specific post translational modification does not exist. For example, an ORF downstream of the catalytic α-subunit and β-subunit of the NHase encodes an activator protein responsible for the posttranslational SO_2H modification of αCys112 residue such that the NHase become functionally active (Odaka et al., 2001). Both the α-subunit and the β-subunit of Fe-type NHase and Co-type NHase include two oxidized Cys residues in (C*X*LC (SO_2H) SC (SOH)) (Zhou et al., 2009). By virtue of which cysteine-sulfinic acid (αCys-SO_2H) and cysteine-sulfenic acid (αCys-SOH) are formed for the catalysis (Zhou et al., 2009).

- ➢ There is no homology between the α-subunit and the β subunits of the NHase. The NHase have basic stoichiometry of α_1 β_1 M_1 (M, Fe^{2+}/Co^{3+}). The modified cysteine residues a Cys112–SOH and a Cys114–SOH, are hydrogen bonded to two arginine residues of the β subunit during post post translational modifications (Endoa et al., 2001).
- ➢ It is known that, activator proteins are essential for the maturation of both metalloenzymes such as Fe-type NHase and Co-type NHases although, the activator proteins are not directly associated with these metalloenzymes (Okamoto et al., 2010).

b. The hydrogen bonding has been found crucial in deciding the pKa of the catalytic residue and the post translational modifications such as formation of dimer as well as the altered C-terminal sequence contribute to the stability and activity of this psychrophilic enzyme. Therefore, this novel amidase is well suited for activity in cold enviornments like that of Antarctica Dry Valley soils. *Nesterenkonia sp.* AN1 found to have aliphatic amidases capable of hydrolyzing short-chain aliphatic amides which belong to a superfamily of amidases having subunits of 30 to 44 kDa existing as active dimers, tetramers, hexamers, octamers and longer oligomeric spirals but in the active form the enzyme exist as dimer whereas, the aliphatic amidases from *Helicobacter pylori*, *Pseudomonas aeruginosa* and *Geobacillus pallidus* exist as hexamers (Fig 3). Furthermore, this psychrophilic enzyme lack the 35-amino-acid C-terminal sequence as well which is a part of C-terminal of aliphatic amidases of *Geobacillus pallidus* and *Pseudomonas aeruginosa* (Nel et al., 2011).

c. Some amidases such as the PA are synthesized as pre-pro-PA (ppPA) precursor consisting of a secretory signal peptide together with the precursor. The ppPA is proteolytically clevaged to produce one small subunit A (24 kDa) and one large subunit (63 kDa) together with the removal of the signal sequence when it is transported to the periplasm (Ignatova et al., 2000).

d. It is a commom observation that the expression of nitrilase in *E.coli* appears as inclusion bodies or as insoluble precipitate of the subunits of the enzyme but soluble expression of nitrilases in *E.coli* has been observed in *R. rhodococcus sp., Bacillus sp., A. faecalis sp.* The inacapability of expression of nitrilase in *E. coli* could be solved by coexpression of GroESL chaperones so as to confirm appropriate folding of the enzyme (Chauhan et al., 2003). Simillarly, the post translational modifications in the NHase of *Rhodococcus sp.* N-774 are essential for their catalytic activity because the cloned gene of NHase in *E.coli* expressed the subunits of the enzyme as insoluble precipitates in inactive form whereas, when shuttle vector E. *coli-Rhodococcus* having transformed *Rhodococcus sp.* with the shuttle vector pK4 is used, this host vector system produced active NHase (Hashimoto et al., 1992; Prepechalova et al., 2001).

The post translational modifications of the nitrilases gene expression products in *E. coli* expression system occur by virtue of auto-oxidation or with the help of chaperones or other proteins.

The co-expression of GroES-GROEL chaperons also increases the activity of enantioselective NHase of *Rhodococcus erythroolis* AJ270 in absence of enhancer P44k whereas, contrary to this observation in presence of enhancer P44k this protein folding chaperons are unable to increase the activity of NHase (Song et al., 2008). The heterologous gene expression of Fe-type NHase of *C. testosteroni* NI1 *E. coli* as $\alpha_2\beta_2$ tetramer is possible only when GroES and GroEL chaperones are coexpressed in the *E.coli* togather with the respective NHase (Stevens et al., 2003).

The *anhE gene* mutants of cobalt containing acetonitrile hydratase (ANHase) gene cluster *anhA* and *anhB* having genes *anhA, anhE and anhB* in sequence in *Rhodococcus jostii* RHA1 could not grow on acetonitrile but grew on acetamide although, the growh on acetonotrile could be restored by providing *anhE* in *trans* because, *AnhE* gene expression product act as a dimeric metallochaperone to deliver cobalt to ANHase such that *AnhA gene* products could assemble ANHase for their activity and maturation of this metalloenzyme (Okamoto et al., 2010). Furthermore, during post trantional events incorporation of cobalt into L-NHase depends upon the exchang of the α-subunit of the apo-L-NHase and the and the holo-NhlAE because, the former lacks cobalt as well as oxidized Cys residues and the later containing Cys-SO^- as metal ligands and a mediator containing cobalt. Hence NhlE also act as by self swapping chapoeron that inserts cobolt into the apo-NhlAE and apo-α-subunit. Moreover, NhlE also act as a metallochaperone in a redox reaction where, non-oxidized cysteine a residue in apo-NhlAE is post-translationally oxidized after cobalt insertion hence is necessary for functional L-NHase maturation. The sAmiD *N*-acetylmuramoyl-l-alanine amidase is a zinc metalloenzyme that forms a dimer by domain swapping (Kiziak et al., 2005; Zhou et al., 2009).

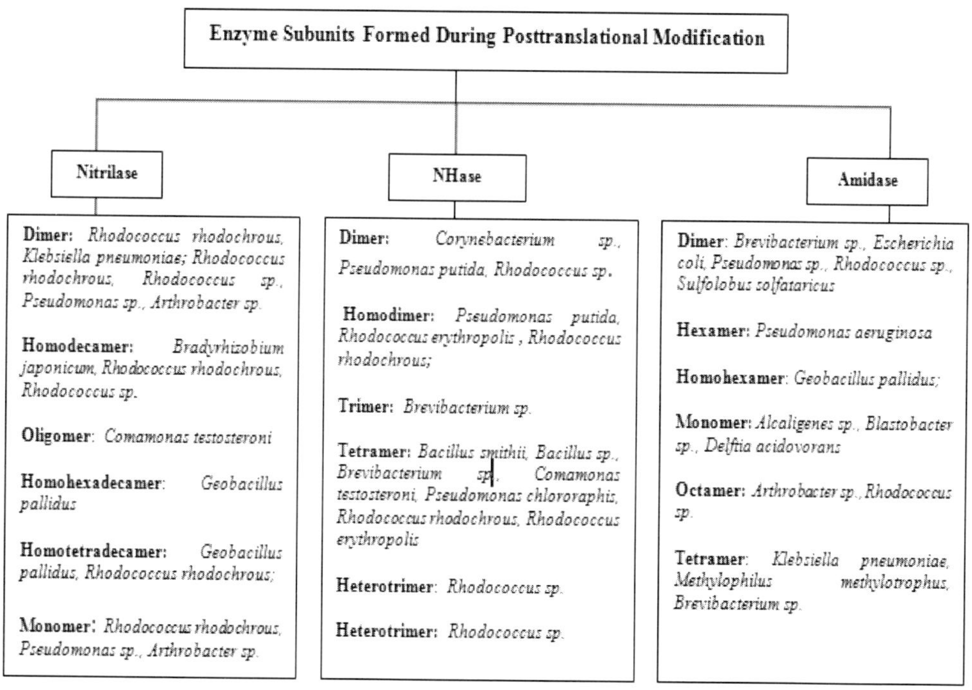

Figure 3. Enzyme subunits formed in nitrile metbolizing enymes in Bacteria.

The physiological role of AnhE as suggested by Kiziak et al. (2009) facilitates the binding of cobalt with the α-subunit of ANHase so as to deliver cobalt ion to ANHase.

Conclusion

In general, bacteria contain both constitutive and/or inducible nitrile metabolizing enzymes controlled by various inducers, enhancers and repressors region of the gene. The structural organizations of genes expressing nitrile hydaratases/amidases are different from nitrilases in bacteria. Hence, different gene products are formed the nitrilase that consists of a single folded polypeptide which forms aggregate so as to form the active nitrilase.The NHases which are heterodimeric (αβ) metalloproteins forming αβ dimers or $\alpha_2\beta_2$ tetramers and the amidases which display the characteristic α-β-β-α fold despite of low sequence conservation are appropriately folded by GroES-GROEL chaperones to gain native conformations of the described nitrile metabolizing enzymes. Some metal ions play an important role in enhancing the folding and stabilization of the subunit polypeptides whereas others are constituents of metalloproteins e.g NHases. Moreover, it has been observed by virtue of mutant variants that the desirable alterations in the structure can alter the substrate specificity and enantioselectivity. In lieu of it, most important are the mutations/alterations in the consensus sequences and motifs present in active site because they have the key role in the catalytic activity of the enzyme.

References

Banerjee, A. Sharma, R. & Banerjee, U.C. (2002). The nitrile-degrading enzymes: current status and future prospects. *Applied Microbiology and Biotechnology, 60,* 33–44.

Brandão, P.F.B. Clapp, J.P. & Bull, A.T. (2003). Diversity of nitrile hydratase and amidase enzyme genes in *Rhodococcus erythropolis* recovered from geographically distinct habitats. *Applied Environmental* Microbiology, *69,* 5754-5766.

Chauhan, S. Wu, S. Blumerman, S. Fallon, R.D. Gavagan, J.E. et al. (2003). Purification, cloning, sequencing and over-expression in *Escherichia coli* of a regioselective aliphatic nitrilase from *Acidovorax facilis* 72W. *Applied Microbiology and Biotechnology, 61,* 118–122.

D'Abusco, A.S. Casadio, R. Tasco, G. Giangiacomo, L. Giartosio, A. et al. (2005). Oligomerization of *Sulfolobus solfataricus* signature amidase is promoted by acidic pH and high temperature. Archaea, *1,* 411-423.

Duran, R. Nishiyama, M. Horinouchi, S. & Beppu, T. (1993). Characterization of nitrile hydratase genes cloned by DNA screening from *Rhodococcus erythropolis. Bioscience, Biotechnology, and Biochemistry,* 57, 1323-1328.

Endoa, I. Nojiria, M. Tsujimuraa, M. Nakasakoc, M. and Nagashima, S. et al. (2001). Fe-type nitrile hydratase. *Journal of Inorganic Biochemistry, 83,* 247–253.

Estepa, J. Luque-Almagro, V.M. Manso, I. Escribano, M.P. Martínez-Luque, M. et al. (2012). The nitlC gene cluster of Pseudomonas pseudoalcaligenes CECT5344 involved in assimilation of nitriles is essential for growth on cyanide. *Environmental Microbiology Reports, 4,* 326–334.

Feng, Y.S. Chen, P.C. Wenc, F.S. Hsiao, W.Y. & Lee, C.M. (2008). Nitrile hydratase from *Mesorhizobium sp.* F28 and its potential for nitrile biotransformation. *Process Biochemistry, 43,* 1391–1397.

Foerstner,K.U. Doerks,T. Muller,J. Raes,J. & Bork,P. (2008). A nitrile hydratase in the eukaryote *Monosiga brevicollis*. *Plos one, 3,* 1-5 e3976.

Galadari, S. Wu, B.X. Mao, C. Roddy, P. El Bawab, S. & Hannun Y.A. (2006). Identification of a novel amidase motif in neutral ceramidase. Biochemical Journal *1,* 687-695.

Hashimoto, Y. Nishiyama, M., Yu, F., Watanabe, I., Horinouchi, S. & Beppu T. (1992). Development of a host-vector system in a *Rhodococcus* strain and its use for expression of the cloned nitrile hydratase gene cluster. *Journal of General Microbiology 138,* 1003-1010.

Heinemann, U. Engels, D. Bürger,S. Kiziak, C. Mattes, R. & Stolz, A. (2003). Cloning of a nitrilase gene from the cyanobacterium *Synechocystis* sp. strain PCC6803 and heterologous expression and characterization of the encoded protein. *Applied and Environmental Microbiology, 69,* 4359-4366.

Herai, S. Hashimoto, Y. Higashibata,H. Maseda,H. & Ikeda, H. (2004). Hyper-inducible expression system for streptomycetes. *Proceedings of the national academy of sciences. USA, 101,* 14031-14035.

Holtze, M.S. Sørensen, J. Hansen, H.C.B. & Aamand, J. (2006). Transformation of the herbicide 2,6-dichlorobenzonitrile to the persistent metabolite 2,6-dichlorobenzamide (BAM) by soil bacteria known to harbour nitrile hydratase or nitrilase. *Biodegradation, 17,* 503-510.

Hong, G.E. Kim, D.G. Bae, J.Y. Ahn, S.H. Bai, S.C et al. (2007). Species-specific PCR detection of the fish pathogen, *Vibrio anguillarum*, using the *amiB* gene, which encodes *N*-acetylmuramoyl-l-alanine amidase. *FEMS Microbiology Letters, 269*, 201–206.

Hongpattarakere, T. Komeda, H. & Asano Y. (2005). Purification, characterization, gene cloning and nucleotide sequencing of D-stereospecific amino acid amidase from soil bacterium: *Delftia acidovorans. Journal of Industrial Microbiology and Biotechnology, 32,* 567–576.

Ignatova, Z. Enfors, S. Hobbie, M. Taruttis, S. Vogt, C. & Kasche V V. (2000). The relative importance of intracellular proteolysis and transport on the yield of the periplasmic enzyme penicillin amiase in *Escherichia coli. Enzyme and Microbial Technology 26,* 165–170.

Ignatova, Z. Mahsunah, A. Georgieva, M. & Kasche, V. (2003). Improvement of posttranslational bottlenecks in the production of penicillin amidase in recombinant *Escherichia coli* strains. Applied Environmental Microbiology, *69,* 1237-1245.

Jin, L.Q. Liu, Z.Q. Zheng, Y.G. & Shen, Y.C. (2010). Identification and characterization of *Serratia marcescens* ZJB-09104, a nitrile-converting bacterium. *World Journal of Microbiology and Biotechnology, 26,* 817–823.

Kaplan, O. Vejvoda, V. Plíhal, O. Pompach, P. Kavan, D. et al. (2006). Biotechnologically relevant enzymes and proteins. purification and characterization of a nitrilase from *Aspergillus niger* K10. *Applied Microbiology and Biotechnology 73,* 567-575.

Kaplan, O. Bezouska,O. Plihal,O. Ettrich,R. Kuliket,N. et al. (2011). Heterologous expression, purification and characterization of nitrilase from *Aspergillus niger* K10. *BMC Biotechnology,* ***11*****,** 1-15.

Kato, Y. Yoshida, S. & Asano, Y. (2005). Polymerase chain reaction for identification of aldoxime dehydratase in aldoxime- or nitrile-degrading microorganisms nitrilase indeed resulted in the aspired islation of a novel type of nitrilase. *FEMS Microbiology Letters, 246*, 243–249.

Kaul, P. & Banerjee, U.C. (2008). Predicting enzyme behavior in nonconventional media: correlating nitrilase function with solvent properties Biomedical and Life Sciences. *Journal of Industrial Microbiology and Biotechnology, 35,* 713-720.

Kiziak, C. & Stolz, A. (2009). Identification of amino acid residues responsible for the enantioselectivity and amide formation capacity of the arylacetonitrilase from *Pseudomonas fluorescens* EBC191. *Applied and Enviornmental Microbiology, 75,* 5592–5599.

Kiziak, C. Conradt, D. Stolz, A. Mattes, R. & Klein, J. (2005). Nitrilase from *Pseudomonas fluorescens* EBC191: cloning and heterologous expression of the gene and biochemical characterization of the recombinant enzyme. *Microbiology,* 151, 3639-3648.

Kiziak, C. Klein, J. & Stolz, A. (2007). Influence of different carboxy-terminal mutations on the substrate-, reaction- and enantiospecificity of the arylacetonitrilase from *Pseudomonas fluorescens* EBC191. *Protein Engineering, Design and Selection, 20,* 385-396.

Kobayashi, M. Komeda, H. Yanaka, N. Nagasawa, T. & Yamada, H. (1992a). Nitrilase from *Rhodococcus rhodochrous* J1. sequencing and overexpression of the gene and identification of an essential cysteine residue. *The Journal of Biological Chemistry, 267,* 20746-20751.

Kobayashi M, Nagasawa, T. & Yamada, H. (1992b). Enzymatic synthesis of acrylamide: a success story not yet over. *Trends in Biotechnology 10,* 402-408.

Kobayashi, M. Izui, H. Nagasawat, T. & Yamada, H. (1993). Nitrilase in biosynthesis of the plant hormone indole-3-acetic acid from indole-3-acetonitrile: Cloning of the *Alcaligenes* gene and site-directed mutagenesis of cysteine residues (nitrile/nucleotide sequence/auxin/gene expression). *Proceedings of the national academy of sciences. USA, Microbiology, 90,* 247-251.

Kobayashi, M. Suzuki, T. Fujita, T. Masuda, M. & Shimizu, S. (1995). Occurrence of enzymes involved in biosynthesis of indole-3-acetic acid from indole-3-acetonitrile in plant-associated bacteria, *Agrobacterium* and *Rhizobium* (nitrile hydratase/purification/auxin/amidase). *Proceedings of the national academy of sciences. USA, Microbiology, 92,* 714-718.

Kolkenbrock, S. Parschat, K. Beermann, B. Hinz, H.J. & Fetzner, S. (2006). *N*-Acetylanthranilate Amidase from *Arthrobacter nitroguajacolicus* Rü61a, an α/β-hydrolase-fold protein active towards aryl-acylamides and -esters, and properties of its cysteine-deficient variant. J. Bacteriol.*188,* 8430-8440.

Komeda, H. Hariyama, N. & Asano,Y. (2006). L-Stereoselective amino acid amidase with broad substrate specificity from *Brevundimonas diminuta*: characterization of a new member of the leucine aminopeptidase family. *Applied Microbiology Biotechnology, 70,* 412–421.

Komeda, H. Kobayashi, M. & Shimizu, S (1996). Characterization of the gene cluster of high-molecular-mass nitrile hydratase (H-NHase) induced by its reaction product in *Rhodococcus rhodochrous* Jl. *Proc. Natl. Acad. Sci. USA, 93,* 4267-4272.

Liu, J-F. Zhang, Z-J. Li, A-T. Pan, J. & Xu, J-H. (2011). Biotechnologically relevant enzymes and proteins. significantly enhanced production of recombinant nitrilase by optimization of culture conditions and glycerol feeding. *Applied Microbiology and Biotechnology, 89,* 665-672.

Makhongela, H.S. Glowacka, A.E. Agarkar, V.B. Sewell, B.T. Weber, B. et al. (2007). A novel thermostable nitrilase superfamily amidase from *Geobacillus pallidus* showing acyl transfer activity. *Biotechnologically Relevant Enzymes and Proteins, Applied Microbiology and Biotechnology,* 75, 801-811

Matamá, T. Carneiro, F. Caparrós, C. Gübitz, G.M. & Cavaco-Paulo, A. (2007). Using a nitrilase for the surface modification of acrylic fibres. *Biotechnol. J. 2,* 353–360.

Mauger, J. Nagasawa, T and Yamada, H. (1990) Occurrence of a novel nitrilase, arylacetonitrilase in *Alcaligenes faecalis* JM3. *Archives of Microbiology*, 155, 1-6.

Miller, J .M. & Gray, D.O. (1982). The Utilization of Nitriles and Amides by a *Rhodococcus* Species. *Journal of General Microbiology, 128,* 1803-1809.

Nagasawa, T. Kobayashi, M. and Yamada, H. (1988) Optimum culture conditions for the production of benzonitrilase by *Rhodococcus rhodochrous* J1. *Archives of Microbiology, 150,* 89-94.

Nel, A.J.M. Tuffin, I.M. Sewell, B.T. & Cowan, D.A. (2011). Unique aliphatic amidase from a psychrotrophic and haloalkaliphilic *Nesterenkonia* isolate. *Applied and Enviornmental Microbiology, 77,* 3696–3702.

Nishiyama, M. Horinouchi, S. Yashi, M.K. Nagasawa, T. Yamada, H. et al. (1991). Cloning and characterization of genes responsible for metabolism of nitrile compounds from *Pseudomonas chlororaphis* B23. *Journal of Bacteriology, 173*, 2465-2472.

Nojiria, M. Nakayama, H. Odakaa, M. Yohdad, M. Takioc, K. et al. (2000). Cobalt-substituted Fe-type nitrile hydratase of *Rhodococcus sp.* N-771. *FEBS Letters, 465,* 173-177.

O'Mahony, R. Doran, J. Coffey, L. Cahill, O.J. Black, G.W. et al. (2005). Characterisation of the nitrile hydratase gene clusters of *Rhodococcus erythropolis* strains AJ270 and AJ300 and *Microbacterium sp.* AJ115 indicates horizontal gene transfer and reveals an insertion of IS1166. *Antonie Van Leeuwenhoek*, 87, 221-232.

Odaka, *M.* Tsujimura, *M.* & Endo, I. (2001). Post-translational modifications in nitrile hydratase family. *RIKEN Review No. 4,* 58-60.

Okamoto, S. Petegem, F.V. Patrauchan, M.A. & Eltis, L.D. (2010). AnhE, a metallochaperone involved in the maturation of a cobalt-dependent nitrile hydratase. *Journal of Biological Chemistry*, 285, 25126-25133.

Pandey, A.K. Raman, S. Proff, R. Joshi, S. Kang, C.M. et al. (2009). Nitrile-inducible gene expression in mycobacteria. Tuberculosis (Edinb), *89*, 12-16.

Payne, M.S. Wu, S. Fallon, R.D. Tudor, G. Stieglitz, B. et al. (1997). A stereoselective cobalt-containing nitrile hydratase. *Biochemistry, 36,* 5447-5454.

Pennartz, A. Généreux, C. Parquet, C. Mengin-Lecreulx, D. & Joris, B. (2009). Substrate-induced inactivation of the *Escherichia coli* AmiD *N*-acetylmuramoyl-l-alanine amidase highlights a new strategy to inhibit this class of enzyme. Antimicrobi*al* Agents Chemotherapy, *53*, 2991-2997.

Podar, M. Eads, J.R. and Richardson, T.H. (2005). Evolution of a microbial nitrilase gene family: a comparative and environmental genomics study. *BMC Evolutionary Biology, 5,* 1-13.

Precigou, S. Wieser, M. Pommares, P. Goulas, P. & Duran, R. (2004). *Rhodococcus pyridinovorans* MW3, a bacterium producing a nitrile hydratase. *Biotechnology Letters, 26,* 1379–1384.

Přepechalová, I. Martínková, L. Stolz, A. Ovesná, M. Bezouška, K. Kopecký, J. & Křen, V. (2001). Purification and characterization of the enantioselective nitrile hydratase from *Rhodococcus equi* A4. *Applied microbiology and biotechnology*, *55*, 150-156.

Robertson, D.E. Chaplin, J.A. DeSantis, G. Podar, M. & Madden, M. et al. (2004). Exploring nitrilase sequence space for enantioselective catalysis. Applied Environmental Microbiolology, *70*, 2429-2436.

Sakashita, T. Hashimoto, Y. Oinuma, K.I. & Kobayashi, M. (2008). Transcriptional regulation of the nitrile hydratase gene cluster in *Pseudomonas chlororaphis* B23. *Journal of Bacteriology, 190*, 4210-4217.

Seffernicka, J.L. Samantac, S.K. Louiec, T.M. Wackett, L.P & Subramanianc, M. (2009). Investigative mining of sequence data for novel enzymes: A case study with nitrilases. *Journal of Biotechnology*, xxx–xxx.

Shi, Y. Yu, H. Sun, X. Zhuoling Tian, Z. & Shen, Z. (2004). Cloning of the nitrile hydratase gene from *Nocardia* sp. In *Escherichia coli* and *Pichia pastoris* and its functional expression using site-directed mutagenesis. *Enzyme and Microbial Technology, 35,* 557–562.

Song, L. Yuan, H.J. Coffey, L. Doran, J. Wang, M.X. Qian, S & O'Reilly C (2008). Efficient expression in E. coli of an enantioselective nitrile hydratase from *Rhodococcus erythropolis*. *Biotechnology Letters, 30,* 7557-7762.

Song, L. Wang, M. Shi, J. Xue, Z. Wang, M.X. et al. (2007). High resolution X-ray molecular structure of the nitrile hydratase from *Rhodococcus erythropolis* AJ270 reveals posttranslational oxidation of two cysteines into sulfinic acids and a novel biocatalytic nitrile hydration mechanism. *Biochemical and Biophysical Research Communications, 362,* 319-324.

Sonke, T. Ernste, S. Tandler, R.F. Kaptein, B. Peeters, W.P.H. et al. (2005). l-Selective Amidase with extremely broad substrate specificity from *Ochrobactrum anthropi* NCIMB 40321. Applied Environmental Microbiology, *71*, 7961-7973.

Sosedov, O. Matzer, K. Bürger, S. Kiziak, C. Baum, S. Altenbuchner, J. & Stolz, A. (2009). Construction of Recombinant Escherichia coli Catalysts which Simultaneously Express an (S)-Oxynitrilase and Different Nitrilase Variants for the Synthesis of (S)-Mandelic Acid and (S)-Mandelic Amide from Benzaldehyde and Cyanide. *Advanced Synthesis & Catalysis*, *351*, 1531-1538.

Sosedov,O. Baum,S. Burger,S. Matzer,K. Kiziak,C. et al. (2010). Construction and application of variants of the *Pseudomonas fluorescens* EBC191 Arylacetonitrilase for Increased Production of Acids or Amides. *Applied and Enviornmental Microbiology, 76,* 3668–3674.

Stalker, D. M. & McBRIDE, K. E. (1987). Cloning and expression in Escherichia coli of a Klebsiella ozaenae plasmid-borne gene encoding a nitrilase specific for the herbicide bromoxynil. *Journal of bacteriology*, *169*, 955-960.

Stevens, J.M. Rao Saroja, N. Jaouen, M. Belghazi, M. Schmitter, J.M. et al. (2003). Chaperone-assisted expression, purification, and characterization of recombinant nitrile hydratase NI1 from Comamonas testosteroni. *Protein Expr Purif. 29,* 70-76.

Takashima, Y. Kawabe, T. & Mitsuda, S. (2000). Factors affecting the production of nitrile hydratase by thermophilic *Bacillus smithii* SC- JO51. *Journal of Bioscience and Bioengineering, 89,* 282-284.

Wilson, S.A. & Drew, R.E. (1995). Identification of two new genes in the *Pseudomonas aeruginosa* amidase operon, encoding an ATPase (AmiB) and a putative integral membrane protein (AmiS). *Journal of Biological Chemistry,* 270, 18818-18824.

Wu, S. Fallon, R.D. & Payne, M.S. (1997). Over-production of stereoselective nitrile hydratase from *Pseudomonas putida* 5B in Escherichia coli: activity requires a novel downstream protein. *Applied Microbiology and Biotechnology, 48,* 704-708.

Yang, Y. Biedendieck, R. Wang, W. Gamer, M. Malten, M. et al. (2006). High yield recombinant penicillin G amidase production and export into the growth medium using *Bacillus megaterium. Microbial Cell Factories*, **5**, 36.

Yeom, S.J. Kim, H.J. Lee, J.K. Kim, D.E. & Oh, D.K. (2008). An amino acid at position 142 in nitrilase from *Rhodococcus rhodochrous* ATCC 33278 determines the substrate specificity for aliphatic and aromatic nitriles. Biochemical Journal., *415*, 401-407.

Yu, H.M. Shi, Y. Luob, H. Tian, Z.L. Zhua, Y.Q. et al. (2006). An over expression and high efficient mutation system of a cobalt-containing nitrile hydratase. *Journal of Molecular Catalysis B: Enzymatic, 43,* 80–85.

Zhang, Z.J. Pan, J. Liu, J.F. Xu, J.H. He, Y.C. et al. (2011). Significant enhancement of (*R*)-mandelic acid production by relieving substrate inhibition of recombinant nitrilase in toluene–water biphasic system. *Journal of Biotechnology, 152,* 24–29.

Zhou, Z. Hashimoto, Y. & Kobayashi, M. (2009). Self-subunit swapping chaperone needed for the maturation of multimeric metalloenzyme nitrile hydratase by a subunit exchange mechanism also carries out the oxidation of the metal ligand cysteine residues and insertion of cobalt. *Journal of Biological Chemistry, 284,* 14930-14938.

In: Molecular Biology of Bacteria
Editors: R. Kumar Gaur and H. K. Gautam
ISBN: 978-1-62618-189-2
© 2013 Nova Science Publishers, Inc.

Chapter 4

Multistress-Protective Molecules in Bacteria

*M. K. Chattopadhyay**
Centre for Cellular and Molecular Biology (CSIR),
Hyderabad, A.P, India

Abstract

The ability of bacteria to adapt themselves to stressful environments (hot springs, glaciers, ocean bottom, dried up lakes, acidic and alkaline water, frozen food compartments) is well-documented. Most often they are challenged with more than one stress condition at a time. Consequently, they evolve with some molecules, which confer the ability to withstand more than one stressor. These multistress-protective molecules could be protein, lipid or some low-molecular weight substances. The importance of these molecules in bioremediation and safe preservation of food materials is discussed with relevant examples.

Bacteria in Stressful Conditions

The word stress is a relative term. Any environment that an organism is not habituated to withstand normally is a stress for it. Thus an aerobic environment is a stress for an obligatory anaerobic organism. However in a general sense, we interpret stress as extremities of several environmental parameters (e.g, temperature, pressure, pH, salt concentration, radiation). Processes that sustain life come to a halt in these conditions. Enzymes, the work horses of a cell, cease to act. In some cases, they get denatured. The cell membrane loses the fluidity, at which it can optimally perform its function. The cell dies.

However, some bacteria adapt themselves to extreme natural environments and we come across reports on the occurrence of bacteria in glaciers, thermal vents, dried up lakes, ocean

* Tel : +91-40-27192545, Fax : +91-40-27160591, E-mail : mkc@ccmb.res.in.

bottom or highly acidic or alkaline water bodies. Organisms adapted to these extreme environments are called extremophiles (Rothschild and Mancinelli 2001). They evolve with a number of strategies to cope with the harsh environments. Some enzymes, isolated from the extremophiles, are capable of working in harsh conditions, which are non-permissive for the same enzymes, isolated from organisms in the temperate environments. These extremozymes are useful for various purposes.

Multistress-Protective Molecules

It is well-known that extreme conditions seldom occur alone in natural environments. They occur in various combinations. Some of the places which are very cold (e.g, Antarctica) are also very dry. Consequently, bacteria which occur there have to adapt themselves both to low temperature and desiccation. Similarly bacteria which dwell in the ocean bottom have to adapt themselves to high pressure and high salinity. If they occur in submarine hydrothermal vents, they have to adapt also to high temperature. Some of these multistress-tolerant bacteria evolve with molecules, which enable them to cope with more than one stress factor at a time. These multistress-protectants interlink mechanisms involved in adaptation of the bacteria to diverse stress factors. The interlink between the ability of some bacteria to survive and grow at low temperature and at the same time to withstand some other stress conditions present in the environment, was highlighted earlier (Chattopadhyay 2008). The multstress-protective molecules are of different types. Some of the examples are discussed here.

- *Proteins:* The Htp G protein is a homolog of the mammalian heat shock protein HSP 90. In the cyanobacterial strain *Synechococcus* sp. PCC7942, Htp G was found to be responsible for basal and acquired thermotolerance, acclimation to low temperature and also acclimation to high oxidative stress (Hossain and Nakamoto 2002, 2003) The Yhb O protein belongs to a superfamily containing chaperones, proteases, catalases and also the Parkinson's disease protein DJ-1. The homologs of the gene encoding this protein are present in widely diverse varieties of living systems ranging from bacteria to humans. Studies with a mutant of *E coli* indicated that the YhbO protein was essential for tolerance of the organism to high temperature, high and low pH, high osmolarity, high oxidative stress and uv-radiation (Abdallah 2007). The GicA protein was found to be required for growth of *E coli* at low temperature and also for its radiation tolerance (Mangoli 2001).
- *Lipids:* Some monounsaturated fatty acids (MUFA) were found to be required by the marine bacterium *Photobacterium profundum* for efficient growth at low temperature and high pressure. Growth of a mutant, deficient in the synthesis of MUFA was found to be adversely affected in presence of either of these two stress conditions (Allen et al 1999)
- *Other molecules:* The term chemical chaperones denotes some small-molecular weight chemical substances (glycine betaine, choline, proline, trehalose) which stabilize the native conformation of cellular proteins during cellular stress. The role of glycine betaine in tolerance of the food borne pathogen *Listeria monocytogenes* to low temperature and high salt was demonstrated earlier (Ko and Smith 1994,

Chattopadhyay 2002). Some other chemical chaperones were also found to encourage growth of the same bacterium at low temperature and high salt (Bayles et al 2000). Bacteria present in food appear to acquire these substances from food and remain protected from harsh conditions (low temperature, high salt, organic acid) widely used for food preservations. Glycine betaine was found to protect *E coli* also from thermal stress (Caldas et al 1999). So glycine betaine is a multi-stress protectant in bacteria. Another chemical chaperone proline, having thermoprotective effect on *E coli* (Chattopadhyay et al 2004) was found to offer protection to *E coli* against low temperature and to some lactic acid bacteria against high osmolarity (Chattopadhyay 2007). Possible involvement of polyphosphates, found to accumulate in the acidocalcisome of some bacterial isolates, obtained from the high altitude Andean wetlands, was postulated in the observed tolerance of the organisms to high salt, uv-radiation and the heavy metal arsenic (Seufferheld 2008). The protective role of trehalose against desiccation (Cytryn et al 2007), low temperature (Kandror et al 2002), high temperature and high salinity (Reina-Bueno et al 2012) was also evidenced.

Possible Mechanism Involved in Multistress-Protection

The ability of some molecules to protect bacteria from more than one stress condition might be explained by the fact that diverse stress factors (high temperature, high salt) most often inflict damaging effects on the cellular machinery using the same mechanism. Most of them adversely affect the native conformation of cellular proteins or alter the fluidity of the cell membrane. Consequently anything, which stabilizes cellular proteins or the cell membrane, offers cross-protection against more than one stressor.

It was reported by Chow and Tung (1998) that frequency of survivors in a strain of *E coli* following freeze storage at -80° C for 24 hours was enhanced almost by 10-times when the sample was warmed at 42 ^{o} C for half an hour prior to freeze-storage.

The inductive synthesis of Dna K/ Dna J in the heat-treated sample was also demonstrated. The investigators postulated that it was the chaperoning effect of the heat shock proteins which helped maintain the native structure of the proteins during freeze-storage and significantly contributed to the enhanced survival. The postulation was bolstered by a subsequent observation that the growth rate of an *E coli* strain at 8° C could be increased by 141-times by cloning and expressing two genes (*cpn* 10, *cpn* 60) that encode two chaperone proteins in an Antarctic bacterium (Ferrer et al 2003) . Hence the chaperoning property of some proteins (molecular chaperones) and small molecules (chemical chaperones) appears to have broad-spectrum protective effects on cellular components of some bacteria. Similarly lipid molecules, which stabilize the membrane in stressful environments are cross-protective against more than one stressor. It is worthwhile to mention that synthesis of an anteiso- fatty acid ($C_{15:0}$), essential for the growth of *Listeria monocytogenes* at low temperature, was stimulated when the organism was grown in presence of glycine betaine (Annous et al 1997). So besides protein-stabilizing effect, this chemical chaperone appears to have also some membrane-stabilizing effect. Stress-tolerance is essential for the sustenance of bacteria that occur in stressful natural environments. Hence the ability of some bacteria (*Deinicoccus radiodurans*) to withstand high dose of ionizing radiation appears to be superfluous since high dose of ionizing radiation is not encountered in natural environments.

The significance of radioresistance however becomes obvious when we look into the mechanism of the damaging effect of radiation (shearing of the cellular DNA) which is shared by desiccation. So the radiation-tolerance is actually an offshoot of desiccation-tolerance (Mattimore and Batista 1996). The DNA-repairing enzymes in the desiccation-tolerant bacteria confer both desiccation and radiation-tolerance and thus interlink mechanisms involved in adaptation of the bacterium to desiccation and radiation.

Production of reactive oxygen species (ROS) is enhanced when some bacteria cells are exposed to low temperature (Chattopadhyay 2011). Production of ROS is also stimulated during desiccation and radiation (Fredrickson 2008). Hence cellular antioxidants appear to have a protective role against all these stress factors. Alternative sigma factors in gram-positive and gram-negative bacteria regulate the expression of a wide variety of genes that are induced in response to various types of stress factors. Thus alternative sigma factors play a protective role against a number of environmental stressors (Schmid and Lidstrom, 2002).

The number of alternative sigma factors in an organism is an indicator of the varieties of environmental challenges it is likely to be confronted with during its lifetime. Thus *Mycoplasma genitalium*, which is an intracellular pathogen, has only one sigma factor while the colon bacterium *E coli* has 7 and the soil bacterium *Streptomyces coelicolor* has 60 alternative sigma factors (Osterberg et al 2011).

Importance

The multistress-protectants are useful for various purposes. The adverse effects of human activities on the ecosystem of some remote and desolate places (e.g, polar regions), which were once pristine, has been a matter of serious concern to the scientists. It is well-known that soil in some Antarctic regions is getting contaminated with petroleum hydrocarbon spillage from air crafts. Removal of the contaminated soil is prohibitively expensive. Bioremediation using petroleum-degrading bacteria is also not feasible because of the poor growth rate of the organism in that environment. The growth rate of these degrader organisms could be improved by overproducing multistress-protective molecules in them through genetic manipulation.

The multistress-protectants are important in food industry for a different reason. During processing, food materials are subjected to various agents (hot water, high salt, organic acid, hydrogen peroxide) in order to annihilate or to reduce the number of contaminating bacteria and thus to make the food materials safe for human use. While many bacteria are killed by those treatments, many others survive. Some of them get more sturdy (bacteria exposed to pH 5 become tolerant to pH 3). Some others become cross-tolerant to some other stressor (Cappozi et al 2009).

Significant increase in resistance to ethanol and sodium chloride was observed in *Listeria monocytogenes* following heat shock (Lou and Yousef 1997).

Dehydration-induced cross-tolerance to various physical (dry heat, uv-irradiation) and chemical (ethanol, sodium hypochlorite, bile salts) stress factors was also evidenced in another food-borne pathogen *Salmonella enterica* (Gruzdev et al 2011). If the multistress-protectants involved in such type of cases could be isolated and characterized, it might be possible to design their inhibitors, which would be useful for safe preservation of food materials.

Conclusion

The characteristics of the multistress-protective molecules of bacterial origin are only beginning to be understood. Some time back, the Department of Biotechnology (Government of India) sanctioned a project proposal submitted by the Centre for Cellular and Molecular Biology (CSIR), Hyderabad for studies on the multistress –protectants in some cold-tolerant Antarctic bacteria, isolated and characterized earlier at this laboratory (Jyotsna and Chattopadhyay 2005, Shivaji and Reddy 2010).

Investigations on multistress- protectants obtained from some other bacteria are likely to provide insights into the mechanism of bacterial adaptation to environmental stress factors.

References

Abdallah J., Caldas, T., Kthiri F., Kern R and Richarme G. (2007), YhbO protects cells against multiple stresses,. *J Bacteriol*, 189: 9140- 9144.

Allen E E., Facciotti D. and Bartlett D.H. (1999). Monounsaturated but not polyunsaturated fatty acids are required for growth of the deep-sea bacterium *Photobacterium profundum* SS9 at high pressure and low temperature. *Appl. Environ. Microbiol*, 65:1710 –1720.

Annous, B.A., Becker L.A., Bayles, D.A., Labeda, D.P and Wilkinson, B.J. (1997) Critical role of anteiso-C $_{15:0}$ fatty acid in the growth of Listeria monocytogenes at low temperatures. *Appl. Environ. Microbiol*, 63: 3887-3894.

Bayles, D. O and Wilkinson, B. J. (2000). Osmoprotectants and cryoprotectants for *Listeria monocytogenes* . *Lett. Appl. Microbiol*, 30 :23-27.

Caldas, T., Demont-Caulet N., Ghazi A. and Richarme, G (1999) Thermoprotection by glycine betaine and choline. *Microbiology*, 145 : 2543-2548.

Capozzi, V., Fiocco D., Amodio M. L., Gallone, A. and Spano, G .(2009), Bacterial stressors in minimally processed food, *Int. J. Mol. Sci.*, 10 : 3076 – 3105.

Chattopadhyay, M.K. (2002). The cryoprotective effects of glycine betaine on bacteria .*Trends. Microbiol*, 10: 311.

Chattopadhyay M.K, Kern R., Mistou M-Y., Dandekar A M, Uratsu S.L. and Richarme, G.(2004). The chemical chaperone proline relieves the thermosensitivity of a *dna K* deletion mutant at 42 degrees C. *J. Bacteriol.*, 186 : 8149-8152.

Chattopadhyay M. K. (2007) Stress adaptation of bacteria: Role of proline. *J. Pur. Appl. Microbiol.* , 1 :339-340 .

Chattopadhyay M K. (2008). Cryotolerance in bacteria: Interlink with adaptation to other stress factors. *Trends Microbiol.* , 16 :455.

Chattopadhyay, M.K,. Raghu, G., Sharma, Y.V.R.K., Biju, A.R., Rajasekharan, M.V.R. and Shivaji, S. (2011). Increase in oxidative stress at low temperature in an Antarctic bacterium. *Curr. Microbiol.,* 62: 544-546.

Chow, K.C. and Tung, W. L. (1998) Overexpression of *dna K/ dna J* and *gro EL* confers freeze tolerance to *Escherichia coli. Biochem. Biophys. Res. Commun.,* 253: 502-505.

Cytron, E.J., Sangurdekar, D.P., Streeter J.G., Frank, W.L., Chang, W.S., Stacey G., Emerich D.W., Joshi, T., Xu D and Shadowsky, M.J. (2007) Transcriptional and physiological responses of *Bradyrhizobium japonicum* to desiccation induced stress. *J. Bacteriol.,* 189: 6751-6762.

Ferrer, M., Chernikova, T. N ., Yakimov, M.M . , Golyshin, P. N. and Timmis, K.N. (2003) Chaperonins govern growth of *Escherichia coli* at low temperatures. *Nat. Biotechnol.,* 21: 1266- 1267.

Fredrickson, J. K., Li, S.W., Gaidamakova, E. K ., Matrosova, V Y., Zhai, M., Sulloway, H M ., Scholten, J. C., Brown, M. G., Balkwill, D. L and Daly, M. J. (2008) Protein oxidation: Key to bacterial desiccation resistance? *ISME. J.* , 2 : 393-403.

Gruzdev N., Pinto R and Sela, S. (2011) Effect of desiccation on tolerance of Salmonella enterica to multiple stresses. *Appl. Environ. Microbiol.*, 77: 1667-1673.

Hossain, M M. and Nakamoto, H. (2002). HtpG plays a role in cold acclimation in cyanobacteria. *Curr Microbiol*, 44 : 291- 296.

Hossain, M.M. and Nakamoto, H. (2003), Role for the cyanobacterial HtpG in protection from oxidative stress. *Curr. Microbiol.*, 46 :70-76.

Jyotsna Ch and Chattopadhyay, M.K. (2005) Antarctic psychrophiles. *Biosci. Biotechnol. Res. Asia* 3: 325-328.

Kandror O., DeLeon A. and Goldberg A.L .(2002). Trehalose synthesis is induced upon exposure of *Escherichia coli* to cold and is essential for viability at low temperatures. *Proc Natl Acd Sci USA,* 99: 9727 – 9732.

Ko R, Smith L. T and Smith GM .(1994). Glycine betaine confers enhanced osmotolerance and cryotolerance on *Listeria monocytogenes. J. Bacteriol* , 176 : 426-431.

Lou, Y and Yousef, A.E (1997) Adaptation to sublethal environmental stress protects *Listeria monocytogenes* against lethal preservation factors. *Appl. Environ.Microbiol* 63 :1252-1255.

Mangoli, S, Sanzgiri, V R. and Mahajan, S.K. (2001) A common regulator of cold and radiation response in *Escherichia coli. J. Environ Pathol. Toxicol. Oncol.*, 20 : 23-26.

Mattimore, V and Battista J.R. (1996) Radioresistance of Deinococcus radiodurans: functions necessary to survive ionizing radiation are also necessary to survive prolonged desiccation. *J. .Bacteriol.*, 178: 633-637.

Osterberg, S., del Peso-Santos T and Shingler V. (2011) Regulation of alternative sigma factor use. *Ann. Rev. Microbiol.*, 65: 37-55.

Reina-Bueno, M., Argandona, M., Salvador, M., Rodriguez-Moya J., Iglesias-Guerra F., Csonka, l.N., Nieto, J.J and Vargas, C. (2012) Role of trehalose in salinity and temperature tolerance in the model halophilic bacterium *Chromohalobacter salixigens. PloS One,* 7: e33587.

Rothschild, L.J and Mancinelli, R. .L (2001) Life in extreme environment. *Nature*, 409, 1092-1101.

Schmid A K. and Lidstrom M. E.(2002). Involvement of two putative alternative sigma factors in stress response of the radioresistant bacterium *Deinococcus radiodurans*. *J. Bacteriol.*, 184,: 6182-6189.

Seufferheld, M J., Alvarez, H.M. and Farias, M. E. (2008). Role of polyphosphates in microbial adaptation to extreme environments. *Appl. Environ. Microbiol.,* 74,: 5867-5874.

Shivaji S and Reddy G.S.N. (2010) Bacterial biodiversity of Antarctica: conventional polyphasic and r RNA approaches. In: Bej AK, Aislabie J, Atlas RM, eds. *Polar microbiology: the ecology, biodiversity and bioremediation potential of microorganisms in extremely cold environments.* CRC Press, pp: 61-93.

In: Molecular Biology of Bacteria
Editors: R. Kumar Gaur and H. K. Gautam

ISBN: 978-1-62618-189-2
© 2013 Nova Science Publishers, Inc.

Chapter 5

Food Microbiome Research: New Insights into Function

Anamika Jha[1], Nisha Daxini[1] and Hemant K. Gautam[2*]
[1]Ashok and Rita Patel Institute of Integrated Study and Research in Biotechnology and Allied Sciences, New Vallabh Vidyanagar, Anand (Gujarat), India
[2]CSIR- Institute of Genomics and Integrative Biology, University Campus, Delhi, India

Abstract

Food microbiome is one of the most rapidly growing analytical approaches that aims at the comprehensive analysis of a large number of metabolites. It is currently applied in numerous fields of research, from discovery of new microbes from food and their role in human health. It is most extensively utilized in the food science field, particularly for food possesses the unique feature of having a large number of natural products and microbial flora that do not exist in any other life form. The current estimation is that over 3,000 food types exist on the earth; each individual food is host to the number of microbes. These food microbes participate in the interaction of the human microflora inside the human body and serve multiple tasks in relation to human health. Therefore, metabolomics analysis of plant based foodstuff like fruit, and cereals, currently gains a great interest. In this article, we briefly described the importance of food microbiome for human health and various analytical techniques which are most useful in metabolomics studies.

Introduction

In recent years, disease research has moved from classical epidemiology and physiology to molecular biology and genomics. Plants and fruits have emerged as a novel and multidisciplinary research field in disease science that aims to elucidate how plants and foods

[*] Corresponding author: hemant@igib.res.in.

can influence human health. It is a well known fact that bioactive compounds from plants and food can interact with genes affecting transcription factors, protein expression and metabolite production. The indigenous microbiota of fruits and vegetables are relatively unexplored and are potential sources of novel natural products for exploitation in medicine, agriculture and industry. It is generally believed that the native microflora of fresh fruits and vegetables are very different from those found on food products of animal origins such as beef, pork and poultry (Liao and Fett, 2001). The native microflora naturally present on the surfaces of fresh produce are assumed to play an important role in maintaining the quality and safety of fresh and fresh-cut fruits and vegetables (Nguyen and Carlin, 1994). Some of them are capable of producing antimicrobial compounds (Guinebretiere et al 2000) or activating plant defense mechanisms (Lund, 1982), leading to the suppression of growth of human pathogens on plant food products and thus are considered beneficial. Other native microflora are capable of inducing spoilage or off-flavors of fresh or packaged produce products and are considered deleterious (Lund, 1982). A better understanding and analysis of the indigenous microflora of fresh produce would provide a basis to develop more effective methods for alleviating several harmful human pathogens. Classical approaches of culture methods frequently used the total aerobic plate counts (APC) of animal and dry food products. The appropriateness of using the same procedure to enumerate the indigenous microflora on fresh produce has not yet been carefully evaluated. So far, there is no standard method or procedure available for enumeration of native APC on fresh produce. The previous studies have described different types of culture media including plate count agar, tryptic soya agar and Levure peptone glucose agar. Molecular methods have always complemented classical methods for studying the countless number of unculturable microbes. The study of complex microbial interactions requires the development of advanced analytical approaches combined with bioinformatics. Thus, to carry out these studies Transcriptomics, Proteomics and Metabolomics approaches are employed together with an adequate integration of the information that they provide. Only a few plants and food products have ever been completely studied relative to their inside microbiota and their importance in curing and controlling of various diseases. The representatives reported to so far belong to the genera *Bacteroids*, *Firmicutes*, *Proteobacteria*, *Lactobacillus* Sp., Actinobacteria and several strains of yeast (Liao and Fett, 2001; Nyanga et al., 2007). So there is always a debate between plant extracts or foods (fruits and vegetables) to cure various diseases.

1. Food-Disease Relationship: Assertions and Reasons

Food is a key environmental factor affecting health and the incidence of many chronic diseases. The effects of nutrient deficiencies, microbial flora, and imbalance of macronutrients or toxic concentrations of certain food components on health have been the main topics in food research. On the other hand, other bioactive food constituents (such as certain polyphenols, vitamins, carotenoids and terpenoids) with significant beneficial effects on health promotion and disease prevention by reducing the process of sustained inflammation that accompanies chronic diseases has been well studied. Most foods are composed of diverse constituents, many of which have specific biological activity contributed by the inner microbiota. There is a lack of effective analytical tools that take into account multiple aspects of the biological effects of foods and thus little is known about their microbial functions and the biological processes involved.

Food genomics focuses on the study of the impact of specific microflora on health through the expression of genetic information by the integration of "omics" technologies such as, Metagenomics, Metabolomics, Proteomics and Transcriptomics (Matthew et al., 2012). One of the main interests in food research relates to health and prevention of lifestyle diseases such as diabetes, heart attack, arthritis and even the chronic diseases like cancer are associated with choice of food and human body microbes. Recent work has shown that obesity is associated with a shift in the presence of the dominant phyla of bacteria in the gut, both in humans and other animal models.

A study was conducted on cardiovascular diseases through food which interpreted that these diseases are complex and multifactorial in their origin, involving not only genetic factors but also a number of behavioral and environmental factors such as exposure to certain food components. It has been found that loss of homeostasis and other biochemical factors are responsible for the disease. In line with, there is a necessity for understanding the molecular mechanisms that describe homeostasis at biochemical, cellular and organ levels, associated with a healthy status and how food affects this homeostatic control. This study reveals that specific microbial /molecular biomarkers may be developed that allow early detection of the onset of disease or pre disease state. However, the discovery of such biomarkers is not easy since food–gene interactions are intricate and complex but there is a chance to develop microbial based biomarkers for various diseases.

2. The Science of 'Omics' Explains Microbiome Function

Recently, it has been recognized that understanding the functional properties of food on health requires the study of the mechanisms of food and other bioactive food constituents at the molecular level. This is supported by the increasingly growing number of studies in humans, animals and cell cultures demonstrating that food and other bioactive compounds in food can regulate gene expression in diverse ways. More specifically, foods and other bioactive food constituents have been referred to as signals that are detected by cellular sensor systems and affect the expression of the genome at several levels (mRNA and proteins) and subsequently, the production of metabolites. The development of Genomics, Metagenomics, Transcriptomics, Proteomics and Metabolomics has created extraordinary opportunities for increasing our understanding about biochemical, microbial and molecular mechanisms that underlies the beneficial or adverse effects of certain bioactive food components. The identity of genes that are involved in the pre- stage to the onset of the disease, as possible molecular biomarkers; and the effect of bioactive food constituents on the expression of these biomarkers involved in controlling crucial molecular pathways need to be explored. To investigate all these parameters it is necessary to integrate analytical approach together with Metagenomics, Transcriptomics, Proteomics and Metabolomics that might be ideal for elucidating the effects of novel and existing functional food constituents and their microbial populations.

2.1. Metagenomics

Metagenomics (also referred to as environmental and community genomics) is a youthful field that has barely scratched the surface of the microbiological landscape, but even in its infancy two features are evident. First, all metagenomics is based on the genomic analysis of microbial DNA that is extracted directly from communities in environmental samples, a process that circumvents microbial culturing. Second, metagenomics is conducted on a grand scale. The complexity of the communities from which metagenomic libraries are built presents both an intellectual attraction and a challenge (Matthew et al., 2012). All microbial communities, whether species-rich or species-poor, contain an uneven species representation, making it difficult to obtain DNA from all the members of the community. Genetic variation within species, species richness and unevenness of species all conspire to obscure the total diversity within communities, necessitating construction of large DNA libraries to provide an accurate representation of the community member.

2.2. Metabolomics

A metabolism represents the ultimate phenotype of cells deduced by perturbation of gene expression and modulation of protein function, which are caused by the environment or mutation. Metabolomics play a key role in understanding the cellular system and decoding the function of genes. Metabolite profiling is a fast growing technology and is useful for phenotyping and diagnostic analyses of plant metabolite and bacterial byproducts. The potential of metabolomics technologies in the study of the genetics underlying metabolite accumulation and metabolic regulation is mostly untapped. The first example of the use of metabolite profiling as a diagnostic tool was to determine the mode of action of various herbicides using an approach based on gas chromatography coupled with mass spectrometry (GC-MS).

2.3. Proteomics

The proteome provides a "picture" of the impact of specific bioactive food and diets in certain organisms, tissue or cell at a particular moment. Thus, the importance of Proteomics as a tool to understand the effect of the diet on health has already been recognized, however, up to now the application of Proteomics in food genomics is still not investigated in detail. The main differences when working with proteins is that there is not an amplification methodology for proteins comparable to PCR. Physical and chemical diversity of proteins are also higher than nucleic acids. They differ among individuals, cell types, and within the same cell depending on cell activity and state. In addition, there are hundreds of different types of post-translational modifications (PTMs), which evidently will influence chemical properties and functions of proteins. The a presence of huge dynamic concentration range of proteins in biological fluids or tissues also hinders the work on proteomics.

2.4. Transcriptomics

In transcriptomics the expression process of genomic information, several steps may be regulated by nutrients and other bioactive compounds in food. Consequently, the analysis of changes in mRNA expression by food nutrients and bioactive food constituents is often the first step to study the flow of molecular information from the genome to the proteome and metablome and one of the main goals in food genomics research. For years, the expression of individual genes has been determined by quantification of mRNA with Northern blotting.

This classical technique has gradually been replaced by more sensitive techniques such as real-time PCR. Both techniques, however, can only analyze gene expression for a limited number of candidate genes at a time. This is an important limitation for their application in food genomics research since the analysis of a reduced number of genes may not provide insights about the causative relationship between the bioactive food constituent and its biological effect.

To overcome this situation, rapid DNA sequencing and microarray technology are used to study the quantitative and comprehensive analysis of changes in the MRNA expression level of several genes. All such technologies suffer from limitations and to overcome this situation probably microbiome study seems the best to study the food microbiome in relation to human health. This technology will facilitate the determination of individual microbial requirements based on the genetic makeup of the person as well as association between food and disease. Food genomics will address the issue of personalized medicine through personalized diet. In the future specific food product may be available not to cure disease but also change your moods from depression to romantic.

Recent studies in mice have suggested that levels of serotonin – known as the 'happy hormone' – are regulated by the amount of bacteria in the gut during early life. The children's may be tested at a young age so that food requirement can be used as preventive medicine.

3. Novel Molecular Techniques to Measure Food Health

An understanding of the coupled biological process at the molecular level is fundamental to assessing the condition of the food. Thus, a number of molecular and cellular techniques are currently being used in conjunction with biological and chemical indicators to increase our ability to evaluate food health.

The modern molecular biological techniques, such as fluorescence in situ hybridization (FISH), reverse transcriptase polymerase chain reaction (RT-PCR), denaturing gradient gel electrophoresis (DGGE) and terminal restriction fragment length polymorphism (T-RFLP), have facilitated the analysis of microbial biodiversity and activity. These techniques have been discussed below in brief.

3.1. DNA Quantification

Measurement of DNA following its extraction from fruit may provide a simple and practicable method for estimating the amount of microbial biomass (Girvan et al., 2004). However, further work on correlating DNA measurements with a particular food type is required.

3.2. Fluorescence Microscopy

The abundance of bacteria in fruit, their cell volumes, and the frequencies of dividing cells can be determined by fluorescence microscopy and computerized image analysis (Bloem et al., 1995). Fruit microbial biomass can be estimated by staining with fluorescent dyes such as fluorescein isothiocyanate.

3.3. FISH

Fluorescence *in Situ* Hybridization (FISH) is a direct, cultivation-independent technique using rRNA-targeted oligonucleotide probes that are frequently used for the identification of microorganisms in different samples. While this technique allows selective visualization of bacterial cells of different phylogenetic groups, it also has some limitations, particularly regarding quantitative analysis of complex samples (Moter and Gobel, 2000).

3.4. RNA Measurement

The composition of microbial communities can be estimated by reverse transcriptase polymerase chain reaction (RT-PCR) followed by gel electrophoresis of the amplified cDNA fragments (Duineveld et al., 2001). The analysis of specific mRNAs reflects the expression of the corresponding gene in the sample. Such measurements can also be done by quantitative real-time RT-PCR, which allows the detection and quantification of of mRNAs present in low amounts in environmental samples, including soils (Pfaffl and Hageleit, 2001). However, this method requires previous knowledge of the sequence of the mRNA of interest.

3.5. SIP

Stable Isotope Probing (SIP) is a culture-independent technique that allows the identification of microorganisms directly involved in specific metabolic processes. In this method, labeled nucleic acids that were synthesized during assimilation of an isotopically enriched substrate are isolated and analyzed (Radajewski et al., 2000).

4. Measurement of Genetic and Functional Biodiversity of Fruit Microbiota with Molecular Techniques

4.1. Fruit Microbiome

Fruits are the attention seekers amongst the food and food products in terms of their microbiological wealth. The fruit health cannot be measured directly; it can be inferred by measuring specific nutritional properties and by observing fruit characters like shape, size and skin texture. There is also increased interest in studying fruit microorganisms in their particular environments, as microbial diversity is intimately related to fruit structure and function which leads to study the fruit metagenomics and fruit metabolome. In this regard, significant progress has been made over the last few years in the development of specific biomarkers and macromolecular probes, enabling rapid and reliable measurements of fruit microbial communities and their impact on fruit health and various new metabolic pathways.

Genetic diversity is most commonly studied by analyzing the diversity of genes encoding 16S rRNA (18S rRNA for eukaryotes). These genes occur in all microorganisms and show species dependent variations in their base compositions. Three methods are commonly applied to examine the diversity of 16S (and 18S) rDNA sequences in total DNA extracted from soil microbial communities: denaturing gradient gel electrophoresis (DGGE), temperature gradient gel electrophoresis (TGGE), and terminal restriction fragment length polymorphisms (T-RFLP).

4.1.1. Terminal Restriction Fragment Length Polymorphism

Differentiation of organisms can also be done according to the patterns derived from cleavage of their DNA (Liu et al., 1997). Thus, in T-RFLP, the specific fingerprint of a community is revealed by analyzing the polymorphism of a certain gene. T-RFLP is a high-throughput, reproducible method that allows the semiquantitative analysis of the diversity of a particular gene in a community.

It requires the extraction of DNA from a food sample and its PCR amplification using a fluorescently labeled primer. T-RFLP yields a mixture of amplicons of the same or similar sizes with a fluorescent label at one end. After purification, the amplicon mixture is digested with a restriction enzyme, which generates fragments of different sizes that are separated by gel or capillary electrophoresis. The separated, labeled fragments are then densitometrically detected and a profile based on fragment lengths is generated.

4.1.2. Denaturing Gradient Gel Electrophoresis

Differences in the melting behavior of small DNA fragments (200 to 700 bp) that differ in as little as a single base substitution can be detected by DGGE (Muyzer et al., 1993). The denaturants used are heat (a constant temperature of 60 °C) and a fixed ratio of formamide (ranging from 0 to 40%) and urea (ranging from 0 to 7M). The position in the gradient where a domain of a DNA fragment melts and thus nearly stops migrating is dependent on the nucleotide sequence in the melted region. The benefit of this approach is that a molecular fingerprint of the community structure is generated for each soil. In fact, each band in each lane of the gel theoretically represents a different bacterial species. In addition, this technique enables the excision and subsequent sequencing of bands, allowing species identification using existing databases.

4.1.3. Temperature Gradient Gel Electrophoresis

In contrast to DGGE, the separation of DNA by TGGE (Heuer and Smalla, 1997) does not depend on a chemical gradient of urea but instead on a precisely defined and controllable temperature gradient. This highly reproducible technique has the same advantages as DGGE. By designing species-specific in situ probes that hybridize to identify bacterial sequences, various species can be examined in even greater detail.

4.1.4. Microbial Resilience

The ability to estimate the relative abundance of each species of microorganisms in the sample, using the three techniques described above, has led to the suggestion that the "equitability index" (J) of numbers of individual species is an important estimation of the resilience of a soil. The use of statistical packages such as Phoretix enables quantification of both diversity indices and equitability (Girvan et al., 2004; Girvan et al., 2003).

The advancement of approaches that do not require the establishment of microbial cultures will undoubtedly enhance our knowledge of bioresources and promote the discovery of new microorganisms with unique capacities for bioremediation, soil restoration, and therapeutic applications.

4.2. Fruit Metagenomics

Fruit metagenomics is the culture-independent genomic analysis of all the micro-organisms in a particular fruit inside or outside. Metagenomics is a new and increasingly sophisticated field which is concerned with the direct isolation of DNA from a defined habitat, followed by cloning (in a surrogate host such as a wild type natural host) of the complete genomes of the entire microbial population in that habitat (Langer et al., 2006). The resulting DNA library is then analysed for functions and sequences of interest. Metagenomics can be divided into a sequence-based and function-driven analysis of uncultured microorganisms (Gabor et al., 2007). Functional metagenomics involves screening metagenomic libraries for a particular phenotype, e.g. salt tolerance, antibiotic production or enzyme activity, and then identifying the phylogenetic origin of the cloned DNA (Dinsdale et al., 2008). Sequence-based approaches, on the other hand, involve screening clones for the highly conserved 16S rRNA genes for identification purposes and then sequencing the entire clone to identify other genes of interest, or large-scale sequencing of the complete metagenome to search for phylogenetic anchors in the reconstructed genomes (Riesenfeld et al., 2004) as represented in Figure 1.

4.3. Metatranscriptomics

Metatranscriptomics (or environmental transcriptomics) involves random sequencing of microbial community mRNA. Neither primers nor probes are needed, so there is no need to anticipate important genes before hand and transcripts from microbial assemblages are sequenced with little bias. Further, paralogous sequences which might cross hybridize on a microarray can be distinguished (Handelsman et al., 2007). The approach is particularly amenable to an experimental framework in which gene expression is monitored while a biotic or abiotic parameter is manipulated. Experimental metatranscriptomics is one of the most powerful tools for understanding the timing and regulation of complex microbial processes within communities and consortia, as well as microbial dexterity in response to changing conditions (Parro et al., 2007, Urich et al., 2008).

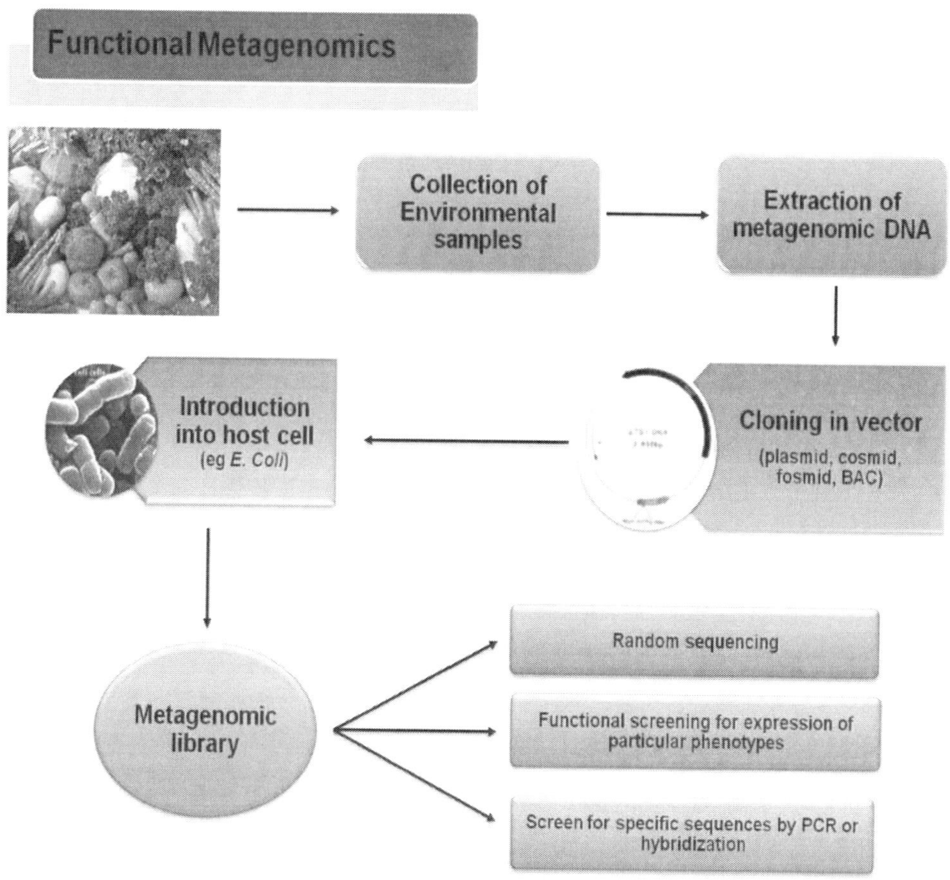

Figure 1. Schematic representation of functional metagenomics.

4.4. Fruit Metabolome in Relation to Functional Genomics

Fruit is an important part of human diet which is having numbers of disease fighting phytochemicals such as flavonoids, glycosides, phenols and phytoestrogens. A number of antioxidant and cancer fighting molecules are isolated from fruits like ellagic acid, ascorbic acid, quercetin, myrecetin and gallic acids (Selmar et al.,1999). A metabolome represent the ultimate phenotype of cells deduced by perturbation of gene expression and modulation of protein function, which are caused by the environment or mutation. Metabolomics can also influence gene expression and protein functions. Metabolomics play a key role in understanding the cellular system and decoding the function of gene. Metabolism will provide better understanding of various metabolic pathways in identification of novel fruit-derived biologically active compounds and genes involved in their biosynthesis which help to engineer fruit with enhanced level of biologically active compounds and fruit quality traits (Fiehn, 2002) as depicted in Figure 2.

During a study conducted on a grass (*Dichanthelium lanuginosum*) that grows around geothermal hot springs at Yellowstone National Park, researchers discovered that if they sterilized the seeds before planting (that is, removed all the associated fungi and bacteria), the grass would no longer grow in the 160 degree Fahrenheit (~70 °C) environment.

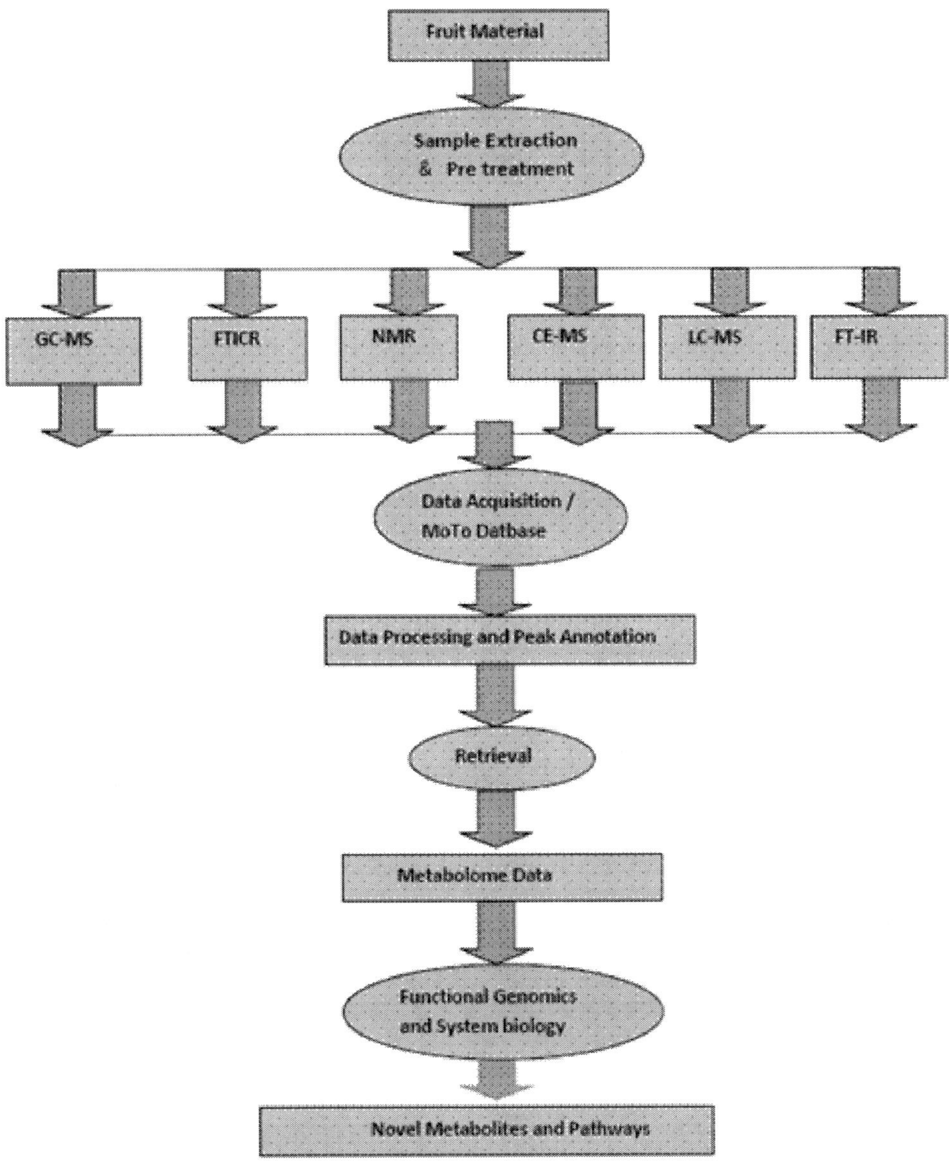

Figure 2. Analytical techniques for fruit metabolomics assay .

In other words, some interaction was going on between the plant and the microbiota present inside and clinging to the seeds that made it possible for this plant to live in such an extreme environment. Further, it was hypothesized that if they took wheat seeds, which normally grow at 100 degrees Fahrenheit, sterilized them, and covered them in bacteria and fungus harvested from the super weeds growing near the geothermal springs, they would grow. Not only did the spore-treated wheat grows at the scorching 160 degree *Fahrenheit* of the geothermal springs, it also required 50 percent less water than normal.

Conclusion

The Human microbiome project has also revealed that the microbes present inside food and humans are responsible for various metabolic activities. This finding has motivated future trends of food research to study how food microbes affect the balance between health and disease by altering the expression of an individual's genetic makeup or how the genetic variability among individuals can influence their predisposition to suffer specific illness related to food as e.g., obesity, diabetics. These aspects have motivated the current trends in food research and this can lead to the development of personalized food to cure various diseases.

References

Bloem, J., Bolhuis, P.R., Veninga, M.R. and Wieringa, J. 1995. Microscopic methods for counting bacteria and fungi in soil. *In*: (eds. Alef, K., Nannipieri, P.) *Methods in applied soil.*

Dinsdale, E.A., Edwards, R.A., Hall, D., Angly, F., Breitbart, M., Brulc, J.M., Furlan, M. and Desnues, C. 2008. Functional metagenomic profiling of nine biomes. *Nature,* 452:629-632.

Duineveld, B.M., Kowalchuk, G.A., Keijzer, A., van Elsas, J.D. and van Veen, J.A. 2001. Analysis of bacterial communities in the rhizosphere of *chrysanthemum* via denaturing gradient gel electrophoresis of PCR-amplified 16S rRNA as well as DNA fragments coding for 16S rRNA. *Appl. Environ. Microbiol.,* 67:172-178.

Fiehn, O. 2002. Metabolomics –the link between genotype and phenotype, *Plant and Mol. Biol.,* 48:155-171.

Girvan, M.S., Bullimore, J., Ball, A.S., Pretty, J.N. and Osborn, A.M. 2004. Responses of active bacterial and fungal communities in soils under winter wheat to different fertilizer and pesticide regimens. *Appl. Environ. Microbiol.,* 70:2692-2701

Girvan, M.S., Bullimore, J., Pretty, J.N., Osborn, A.M. and Ball, A.S. 2003. Soil type is the primary determinant of the composition of the total and active bacterial communities in arable soils. *Appl. Environ. Microbiol.,* 69:1800-180.

Guinebretiere, M.H., Nyuyen-the, C., Morrison, N., Reich, M. and Nicot, P. 2000. Isolation and characterization of antagonists for the biocontrol of postharvest wound pathogen *Botrytis cinerea* on straw- berry fruits. *J. Food Prot.* 63:386-394.

Handelsman, J., Tiedje, J., Alvarez-Cohen, L., Ashburner, M. and Cann, I.K.O. 2007. The New Science of metagenomics: revealing the secrets of our microbial planet. Washington, DC: The National Academies Press.

Heuer, H. and Smalla, K. 1997. Application of denaturing gradient gel electrophoresis and temperature gel electrophoresis for studying soil microbial communities. *In:* (eds. van Elsas, J.D., Trevors, J.T., Wellington, E.M.H.) *Modern soil microbiology*. Marcel Dekker, New York, pp. 353-373.

Langer, M., Gabor, E.M., Liebeton, K., Meurer, G., Niehaus, F., Schulze, R., Eck, J. and Lorenz, P. 2006. Metagenomics: an inexhaustible access to nature's diversity. *Biotechnol. J.,* 1:815-821.

Liao, C.H. and Fett, W.F. 2001. Analysis of native microflora and selection of strains antagonistic to human pathogens on fresh produce. *J. Food Protect.*, 64(8): 1110-1115

Liu, W.T., Marsh, T.L., Cheng, H. and Forney, L.J. 1997. Characterization of microbial diversity by determining terminal restriction fragment length polymorphisms of genes encoding 16S rRNA. *Appl. Environ. Microbiol.,* 63:4516-4522.

Lund, B.M. 1982. The effect of bacteria on post-harvest quality of vegetables and fruits, with particular reference to spoilage. *In* (Rhodes-Roberts M. and Skinner, E.A., eds) *Bacteria and Plants.* Academic Press, NewYork. pp.133-153.

Matthew, B., Scholz, M.B., Lo, C.C. and Chain, P.S.G. 2012. Next generation sequencing and bioinformatic bottlenecks: the current state of metagenomic data analysis. *Curr. Opinion Biotechnol.* 23:9–15.

Moter, A. and Göbel, U.B. 2000. Fluorescence in situ hybridization (FISH) for direct visualization of microorganisms. *J. Microbiol. Methods,* 41:85-112.

Muyzer, G., de Waal, E.C. and Uitterlinden, A.G. 1993. Profiling of complex microbial populations by denaturing gradient gel electrophoresis analysis of polymerase chain reaction-amplified genes coding for 16S ribosomal-RNA. *Appl. Environ. Microbiol.,* 59:695-700.

Nguyen-the, C. and Carlin, F.C. 1994. The microbiology of minimally processed fresh fruits and vegetables. *Crit. Rev. Food Sci. Nutr.* 34: 371-401.

Nyanga, L.K., Nout, M.J., Gadaga, T.H., Theelen, B., Boekhout, T., Zwietering, M.H. (2007). Yeast and Lactic acid bacteria microbiota from masau (Ziziphus mauritiana) fruits and their fermented fruit pulp in Zimbabwe. *Intl. J. Food Microbiol,* 120: 159-166.

Parro, V., Moreno-Paz, M. and Gonzalez-Toril, E. 2007. Analysis of environmental transcriptomes by DNA microarrays. *Env. Microbiol.,* 9:453-464.

Pfaffl, M.W. and Hageleit, M. 2001. Validities of mRNA quantification using recombinant RNA and recombinant DNA external calibration curves in real-time RT-PCR. *Biotechnol. Lett.,* 23:275-282.

Radajewski, S., Ineson, P., Parekh, N.R. and Murrell, J.C. 2000. Stable-isotope probing as a tool in microbial ecology. *Nature,* 403:646-649.

Riesenfeld., C.S., Schoss, P.D. and Handelsman, J. 2004. Metagenomics: genomic analysis of microbial communities. *Annu. Rev. Genet.,* 38:525-552.

Selmar, D. 1999. Cyanide in food: Biology of cyanide glucosides and related nutritional problems. *In*: (ed. Romeo, J.T.), *Phytochemicals in human health protection, nutrition and plant defense.* Kluwer Academic Publishing, New York. pp. 369-392.

Urich, T., Lanze´n A., Qi J., Huson, D.H. and Schleper C., et al., 2008. Simultaneous Assessment of Soil Microbial Community Structure and Function through Analysis of the Meta-Transcriptome. *PLoS ONE* 3: e2527. doi:10.1371/ Journal.

In: Molecular Biology of Bacteria
Editors: R. Kumar Gaur and H. K. Gautam
ISBN: 978-1-62618-189-2
© 2013 Nova Science Publishers, Inc.

Chapter 6

Microorganism and Key Degradative Enzymes: Tools of Remediation of Heavy Metals and PAHs

Pankaj Kumar Jain[1*], Vivek Bajpai[2] and Madan Lowry[3]
[1]Biotechnology Laboratory, Department of Science, MITS University, Lakshmangarh, Rajasthan, India
[2]Department of Microbiology, ITM University, Gwalior, MP, India
[3]Center for Advance Studies, Department of Zoology, University of Rajasthan, Rajasthan, India

Abstract

Pollution is increasing due to the rapid industrialization on all over the world. Heavy metals and Poly cyclic hydrocarbons (PAHs) pollution is nowadays one of the most important environmental concerns. The modes, through pollutants enter in the biosphere is that of industrial effluents. These pollution problems often result in huge disturbances of both the biotic and abiotic components of the ecosystem. The currently used physical and chemical treatments are effective for the degradation of petroleum products and heavy metals but they lag behind in the desired properties, apart they frequently produce many hazardous compounds which are potent immunotoxicants and carcinogenic for living beings. Bioremediation of petroleum oil and heavy metal contaminated site is accomplished with the help of a diverse group of microorganisms (particularly the indigenous bacteria present in soil) and plants. It uses relatively low-cost, low technique, which generally have a high public acceptance and can often be carried out on site. It is effective treatment in terms of efficacy, safety on long terms use, cost and simplicity of administration. The present review gives an idea about the mycoremediation, phytoremediation, bioaugmation, biostimulation, landfarming, bioventing, bioleaching, composting process etc. and enzymes used in bioremediation process.

[*] Corresponding author's Email adress: pankajbiotech2001@yahoo.com.

Keywords: Mycoremediation, phytoremediation, bioaugmation, PAHs, enzymes of bioremediation

Introduction

A variety of heavy metals and Poly cyclic hydrocarbons (PAHs) are discharged in the environment from a large number of petroleum refineries/industries.

The contaminated soil, groundwater, and/or wastewater contain a mixture of contaminant types, often at widely varying concentrations. These may include salts, organics, alcohols, phenols, acid, radionucliods, heavy metals, PAHs, and trace elements like zinc, cadmium, mercury, copper, chromium, lead (Pathak et al., 2010). The simultaneous cleanup of multiple, mixed contaminants using conventional chemical and thermal methods are both technically difficult and expensive; these methods also destroy the biotic component of soils.

Bioremediation is a pollution treatment technology in which bacteria, fungi, and plants degrade pollutants, chemicals through utilization or transformation of the substances. Bioremediation is a process that uses microorganisms or their enzymes to promoted gradation and/or removal of contaminants from the environment. Microorganisms used to perform the function of bioremediation are known as bioremediators. It is the use of biological processes to return the environment to its original state. In other words the goal of bioremediation is to make the environment less toxic. Bioremediation includes use of enzymes, growth stimulants, bacteria, fungi, or plants to degrade, transform, sequester, mobilize, or contain contaminant organics, inorganics, or metals in soil, water, or air. If we accept the ''Doctrine of Infallibility'' i.e., there is no compound known to man that microorganisms cannot degrade (Alexander, 1965), then bioremediation becomes one of the great solutions for our environmental problems. Unfortunately, while the Doctrine of Infallibility may be absolutely true, the rates of biodegradation or transformation of some compounds is so slow as to be negligible for some bioremediation purposes. In addition, the conditions (environmental or biological) that allow certain biological reactions to take place may not be obtainable in many environments (Fewson, 1988).

Table 1. Major pollutants in nature

Organic	Inorganic	Air
Pesticides	Arsenic	Particulates
N and P derivatives	Cadmium	Greenhouse gases
PAHs	Chromium	Smog forming compounds
BTEX	Copper	
PCPs	Lead	
NAPL	Cyanide	
Plastics, Biopolymers		
Dyes		
Phenols, Chlorophenols		
Choroanilines		
Bleach plant effluents		
Nitrocompounds		

Bioremediation processes have been classified in three broad categories, according to place and soil handling/ conditioning: *In-situ, Ad-situ,* and *Ex-situ*. The second and third class of processes are useful for the remediation of-

1. Sludge, soils or sediments polluted with high concentration of recalcitrant contaminants (Zhang, 2000), for instance polynuclear aromatic hydrocarbons (Oberbremer et al., 1990; Pinelli et al, 1997; Wang and Vipulanandar, 2001; Janikowski et al., 2002), diesel (Nano et al., 2003; Fava et al., 2004), explosives (Fuller and Manning, 2004), pesticides and chlorinated organic pollutants (Robles-González et al., 2003), oily sludges from the petrochemical industry (Castaldi and Ford, 1992).
2. Clay and stratified soils with low hydraulic conductivity and low permeability accompanied with high contents of organic matter (Robles-González et al., 2003; Woo and Park, 1999).
3. Soils in regions and areas where environmental conditions are adverse to biological processes, for instance, low temperature that negatively affects biodegradation rates (Mueller, 1991).

Contaminated sites that require a short remediation time because of regulatory or other pressures (Cookson, 1995; Crawford and Crawford, 1996; Eweis et al., 1998). The microorganisms most responsible for bioremediation are bacteria. When indigenous bacteria degrade compounds under existing subsurface conditions, the degradation is called passive bioremediation or natural attenuation. Natural attenuation is most common in the subsurface, where bacteria are plentiful and where pollutant concentrations are low enough (RTDF, 1996). Where natural bacteria are not enough, fertilizers can apply to increase the bioavailability within the medium (biostimulation). Metal contaminants are commonly found in soils, sediments, and water. Metal pollutants can be produced through industrial processes such as mining, refining, and electroplating. A key factor to the remediation of metals is that metals are non-biodegradable, but can be transformed through sorption, methylation, and complexation, and changes in valence state. So, we can use various microorganisms for the metal bioremediation (Bajpai et al., 2011; Dhakar and Jain, 2011; Tilawat and Jain, 2011). These transformations affect the mobility and bioavailability of metals. At low concentrations, metals can serve as important components in life processes, often serving important functions in enzyme productivity. However, above certain threshold concentrations, metals can become toxic to many species. Fortunately, microorganisms can affect the reactivity and mobility of metals. Microorganisms that affect the reactivity and mobility of metals can be used to detoxify some metals and prevent further metal contamination. Major pollutants in nature are given in Table 1.

Types of Bioremediation

1. Mycoremediation

Mycoremediation is a form of bioremediation in which fungi are used to decontaminate the area and to degrade or sequester contaminants in the environment. The term mycoremediation refers specifically to the use of fungal mycelia in bioremediation. This term was coined by Paul Stamets

One of the primary roles of fungi in the ecosystem is decomposition, which is performed by the mycelium.

The mycelium secretes extracellular enzymes and acids that break down lignin and cellulose, the two main building blocks of plant fiber. These are organic compounds composed of long chains of carbon and hydrogen, structurally similar to many organic pollutants. The key to mycoremediation is determining the right fungal species to target a specific pollutant. In an experiment, a plot of soil contaminated with diesel oil was inoculated with mycelia of oyster mushrooms; traditional bioremediation techniques (bacteria) were used on control plots. After four weeks, more than 95% of many of the PAH (polycyclic aromatic hydrocarbons) had been reduced to non-toxic components in the mycelial-inoculated plots. It appears that the natural microbial community participates with the fungi to break down contaminants, eventually into carbon dioxide and water. Wood-degrading fungi are particularly effective in breaking down aromatic pollutants (toxic components of petroleum), as well as chlorinated compounds (Singh, 2006). Mycofiltration is a similar process, using fungal mycelia to filter toxic waste and microorganisms from water in soil.

2. Phytoremediation

Phytoremediation is an emerging cleanup technology for contaminated soils, groundwater, and wastewater, in term of low-tech and low-cost. It is a type of bioremediation process that uses various types of plants to remove, transfer, stabilize, and/or destroy contaminants in the soil and groundwater. In other words Phytoremediation ('phyto' means plant) is a generic term for the group of technologies that use plants for remediating soils, sludges, sediments and water contaminated with organic and inorganic contaminants. Phytoremediation process is the efficient use of plants to remove, detoxify or immobilize environmental contaminants in a growth matrix (soil, water or sediments) through the natural biological, chemical or physical activities and processes of the plants.

Phytoremediation is the engineered use of green plants, including grasses, forbs, and woody species, to remove, contain, or render harmless such environmental contaminants as heavy metals, trace elements, organic compounds, and radioactive compounds in soil or water. This definition includes all plant-influenced biological, chemical, and physical processes that aid in the uptake, sequestration, degradation, and metabolism of contaminants, either by plants or by the free-living organisms that constitute the plant's rhizosphere. Phytoremediation takes advantage of the unique and selective uptake capabilities of plant root systems, together with the translocation, bioaccumulation, and contaminant storage/degradation abilities of the entire plant body. Plant-based soil remediation systems can be viewed as biological, solar-driven, pump-and-treat systems with an extensive, self extending uptake network (the root system) that enhances the below-ground ecosystem for subsequent productive use.

This technology has been receiving attention lately as an innovative, cost-effective alternative to the more established treatment methods used at hazardous waste sites.

The U.S. Environmental Protection Agency (EPA) seeks to protect human health and the environment from risks associated with hazardous waste sites, while encouraging development of innovative technologies such as phytoremediation to more efficiently clean up these sites. For treating contaminated wastewater, the phytoremediation plants are grown in a bed of inert granular substrate, such as sand or pea gravel, using hydroponic or aeroponic techniques.

The wastewater, supplemented with nutrients if necessary, trickles through this bed, which is ramified with plant roots that function as a biological filter and a contaminant uptake system. An added advantage of phytoremediation of wastewater is the considerable volume reduction attained through evapotranspiration (Hinchman and Negri, 1994). Phytoremediation may be applied wherever the soil or static water environment has become polluted or is suffering ongoing chronic pollution.

There are several different types of phytoremediation mechanisms. These are-

i. *Rhizosphere biodegradation- In* Rhizosphere biodegradation plant releases natural substances through its roots, supplying nutrients to microorganisms in the soil. Microorganisms enhance biological degradation.
ii. *Phyto-stabilization-* This process based on chemical compounds produced by the plant which immobilize contaminants, rather than degrade them.
iii. *Phyto-accumulation (phyto-extraction)-* Plant roots sorb the contaminants along with other nutrients and water. The contaminant mass is not destroyed but ends up in the plant shoots and leaves. This method is used primarily for wastes containing metals. At one demonstration site, water-soluble metals are taken up by plant species selected for their ability to take up large quantities of lead (Pb). The metals are stored in the plants aerial shoots, which are harvested and either smelted for potential metal recycling / recovery or are disposed of as a hazardous waste. As a general rule, readily bioavailable metals for plant uptake include cadmium, nickel, zinc, arsenic, selenium, and copper. Moderately bioavailable metals are cobalt, manganese, and iron. Lead, chromium, and uranium are not very bioavailable. Lead can be made much more bioavailable by the addition of chelating agents to soils. Similarly, the availability of uranium and radio-cesium 137 can be enhanced using citric acid and ammonium nitrate, respectively.
iv. *Hydroponic Systems for Treating Water Streams (Rhizofiltration)-* Rhizofiltration is similar to phyto-accumulation, but the plants used for cleanup are raised in greenhouses with their roots in water. This system can be used for *ex-situ* groundwater treatment. That is, groundwater is pumped to the surface to irrigate these plants. Typically hydroponic systems utilize an artificial soil medium, such as sand mixed with perlite or vermiculite. As the roots become saturated with contaminants, they are harvested and disposed of.
v. *Phyto-volatilization-* In this process, plants take up water containing organic contaminants and release the contaminants into the air through their leaves.
vi. *Phyto-degradation-* In this process, plants actually metabolize and destroy contaminants within plant tissues.
vii. *Hydraulic Control-* In this process, trees indirectly remediate by controlling groundwater movement. Trees act as natural pumps when their roots reach down towards the water table and establish a dense root mass that take up large quantities of water. A poplar tree, for example, pulls out of the ground 30 gallons of water per day and a cottonwood can absorb up to 350 gallons per day.

3. Bioaugmentation

Bioaugmentation is the application of specifically selected bacteria (microbes) into a pollutant treatment system to enhance the efficiency of bioremediation process. Bioaugmentation has many applications in all the types of bioremediation. Introducing specialized microbes at a site to essentially "eat" the contamination is called bioaugmentation. It has proven to be a fast, effective and affordable remediation alternative, and it is finding favor among site managers and remediation experts. Bioaugmentation is a powerful tool that can be used as a stand-alone remediation alternative, but it works quite well with many other remediation technologies to expedite site clean-ups.

It is the introduction of a group of natural microbial strains or a genetically engineered variant to treat contaminated soil or water. Usually the steps involve studying the indigenous varieties present in the location to determine if biostimulation is possible. If the indigenous variety do not have the metabolic capability to perform the remediation process, exogenous varieties with such sophisticated pathways are introduced. Bioaugmentation is commonly used in municipal wastewater treatment to restart activated sludge bioreactors. Most cultures available contain a research based consortium of Microbial cultures, containing all necessary microorganisms (*Bacillus licheniformis, Bacillus thurengensis, Bacillus sterothemophilus, Penicillium* sp., *Aspergillus* sp., *Flavobacterium, Arthrobacter, Pseudomonas, Streptomyces, Saccaromyces, Triphoderma* etc.). The activated sludge systems are generally based on microorganisms like bacteria, protozoa, nematodes, rotifers and fungi capable to degrade bio degradable organic matter. It has been utilized in agriculture for many years. The inoculation of legumes with symbiotic, nitrogen-fixing *Rhizobium* spp. dates back to the 1800s. Attempts have also been made to utilize bioaugmentation with free-living or plant-associated nitrogen-fixing bacteria such as *Azotobacter* or *Azospirillum* spp. to increase plant yields. (Monib, 1979; Ramos, 2002).

Like any remediation technology, the speed and success of a bioaugmentation product depends almost entirely on a clear understanding of site-specific conditions. Some bioaugmentation products, such as *Pseudomonas* sp., are better suited for aerobic conditions, while others, like *Dehalococcoides* sp. perform best in anaerobic environments. Bioaugmentation is sometimes confused with biostimulation, which is the addition of chemicals or nutrients into the subsurface to stimulate the indigenous bacteria. The stimulation process often requires some ramp-up time as the bacterial population increases to the numbers necessary to effectively degrade the contaminants. Biostimulation also presents the risk of stimulating a species or strain of microbes that are unable to metabolize the contaminants of concern or favor other food sources, thus stalling remediation efforts.

Bioaugmentation for Enhanced Aerobic Biodegradation

Aerobic Biodegradation is the degradation of petroleum hydrocarbons or any aerobically degradable contaminants in soil and groundwater. It is rare if ever that the addition or augmentation of aerobic degraders is required to facilitate aerobic biodegradation processes.

Research has shown that aerobes are ubiquitous and are typically limited by the availability of oxygen to sustain and or grow their populations.

Bioaugmentation for Enhanced Anaerobic Biodegradation

Anaerobic Biodegradation is related to bioaugmentation for the degradation of chlorinated compounds in soil and ground water; many remediation practitioners today recommend the addition of cultured anaerobic microorganisms for the biodegradation of certain target compounds. The compounds most often targeted with anaerobic degraders include: dichloroethylene (DCE) and vinyl chloride. These two significant subsurface contaminants can be produced in relatively high quantities during the reductive dechlorination of parent compounds such as perchloroethylene (PCE) and trichloroethylene (TCE). Specific strains of anaerobic bacteria have been isolated, cultured and are now commercially available for the biodegradation of DCE and vinyl chloride.

4. Biostimulation

All engineered bioremediation can be characterized as either biostimulation, i.e., the addition of nutrients, or bioaugmentation, i.e., the addition of organisms, or processes that use both. The problems with adding chemical nutrients to sediment and groundwater are fundamentally different from those of adding organisms. Simple infiltration of soil and subsequently groundwater is physically quite different in the two processes (Alfoldi, 1988).

Biostimulation involves the modification of the environment to stimulate existing bacteria capable of bioremediation. This can be done by addition of various forms of rate limiting nutrients and electron acceptors, such as phosphorus, nitrogen, oxygen, or carbon (e.g. in the form of molasses). Additives are usually added to the subsurface through injection wells, although injection well technology for biostimulation purposes is still emerging. Removal of the contaminated material is also an option, albeit an expensive one. Biostimulation can be enhanced by bioaugmentation. This process, overall, is referred to as bioremediation and is an EPA-approved method for reversing the presence of oil or gas spills.

The primary advantage of biostimulation is that bioremediation will be undertaken by already present native microorganisms that are well-suited to the subsurface environment, and are well distributed spatially within the subsurface. The primary disadvantage is that the delivery of additives in a manner that allows the additives to be readily available to subsurface microorganisms is based on the local geology of the subsurface. Tight, impermeable subsurface lithology (tight clays or other fine-grained material) make it difficult to spread additives throughout the affected area. Fractures in the subsurface create preferential pathways in the subsurface which additives preferentially follow, preventing even distribution of additives.

5. Bioventing

Bioventing is an *In-Situ* remediation technology that uses indigenous microorganisms to biodegrade organic constituents adsorbed to soils in the unsaturated zone. Soils in the capillary fringe and the saturated zone are not affected. In Bioventing, the activity of the indigenous bacteria is enhanced by inducing air (or oxygen) flow into the unsaturated zone (using extraction or injection wells) and, if necessary, by adding nutrients. Bioventing primarily assists in the degradation of adsorbed fuel residuals, but also assists in the degradation of volatile organic compounds (VOCs) as vapors move slowly through biologically active soil. The rate of natural degradation is generally limited by the lack of oxygen and other electron acceptor (i.e., a compound that gains electrons during biodegradation) rather than by the lack of nutrients (i.e., electron donors). In conventional bioventing systems, oxygen is delivered by an electric blower to subsurface wells. In contrast to soil vapor vacuum extraction, bioventing uses low airflow rates to provide only enough oxygen to sustain microbial activity.

Passive bioventing systems use natural air exchange to deliver oxygen to the subsurface via bioventing wells. A one-way valve, installed on a vent well, allows air to enter the well when the pressure inside the well is lower than atmospheric pressure. When atmospheric pressure drops (due to a change in barometric pressure) below the subsurface pressure, the valve closes, trapping the air in the well and increasing oxygen to the soil surrounding the well.

Principles of Bioventing

Soil normally contains large numbers of diverse microorganisms including bacteria, algae, fungi, protozoa, and actinomycetes. In well-aerated soils, which are most appropriate for bioventing, these organisms are generally aerobic. Of these organisms, the bacteria are the most numerous and biochemically active group, particularly at low oxygen levels. Bacteria require a carbon source for cell growth and an energy source to sustain metabolic functions required for growth. Nutrients, including nitrogen and phosphorus, are also required for cell growth. Hydrocarbon-degrading aerobic bacteria use oxygen to metabolize organic material to yield carbon dioxide and water, a process commonly referred to as aerobic respiration. To degrade large amounts of petroleum hydrocarbons, a substantial bacterial population are required which, in turn, requires oxygen for both the metabolic process and the growth of the bacterial mass itself. Approximately 3 to 3.5 pounds of oxygen are needed to degrade one pound of petroleum product.

The most important factors that control the effectiveness of bioventing are:

i. The permeability of the petroleum-contaminated soils. This will determine the rate at which oxygen can be supplied to the hydrocarbon-degrading microorganisms found in the subsurface.
ii. The biodegradability of the petroleum constituents. This will determine both the rate at which and the degree to which the constituents will be metabolized by microorganisms.

In general, the type of soil will determine its permeability. Fine-grained soils (*e.g.*, clays and silts) have lower permeabilities than coarse-grained soils (*e.g.*, sands and gravels). The biodegradability of a petroleum product constituent is a measure of its ability to be metabolized by hydrocarbon-degrading bacteria that produce carbon dioxide and water as byproducts of microbial respiration. Petroleum products are generally biodegradable regardless of their molecular weight, as long as indigenous microorganisms have an adequate supply of oxygen and nutrients. For heavier constituents (which are less volatile and less soluble than many lighter components), biodegradation will exceed volatilization as the primary removal mechanism, even though biodegradation is generally slower for heavier constituents than for lighter constituents.

The ability of a soil to transmit air, which is of prime importance to bioventing, is reduced by the presence of soil water, which can block the soil pores and reduce air flow. This is especially important in fine-grained soils, which tend to retain water.

Soil structure and stratification are important to bioventing because they affect how and where soil vapors will flow within the soil matrix when extracted or injected. Structural characteristics such as microfracturing can result in higher permeabilities than expected for certain soils (*e.g.*, clays). Increased flow will occur in the fractured but not in the unfractured media. Stratification of soils with different permeabilities can dramatically increase the lateral flow of soil vapors in more permeable strata while reducing the soil vapor flow through less permeable strata. This preferential flow behavior can lead to ineffective or extended remedial times for less-permeable strata or to the possible spreading of contamination if injection wells are used.

Applications of Bioventing

Bioventing remediates soils contaminated with fuel. Bioventing techniques have been successfully used to remediate soils contaminated with non-chlorinated solvents, some pesticides, wood preservatives, and other organic chemicals. All aerobically biodegradable constituents can be treated by Bioventing. In particular, Bioventing has proven to be very effective in remediating releases of petroleum products including gasoline, jet fuels, kerosene, and diesel fuel.

Advantages of Bioventing

- Uses readily available equipment; easy to install.
- Creates minimal disturbance to site operations. Can be used to address inaccessible areas (*e.g.*, under buildings).
- Requires short treatment times: usually 6 months to 2 years under optimal conditions.
- Cost competitive: $45-140/ton of contaminated soil.
- Easily combinable with other technologies (*e.g.*, air sparging, groundwater extraction).
- May not require costly offgas treatment.

Disadvantages of Bioventing

- High constituent concentrations may initially be toxic to microorganisms.
- Not applicable for certain site conditions (*e.g.*, low soil permeabilities, high clay content, insufficient delineation of subsurface conditions).
- Cannot always achieve very low cleanup standards.
- Permits generally required for nutrient injection wells (if used). (A few states also require permits for air injection.)
- Only treats unsaturated-zone soils; other methods may also be needed to treat saturated-zone soils and groundwater.

6. Bioleaching

Microbial leaching often referred to as bioleaching, involves the microbial catalysed acidification and solubilization of heavy metals and is one of the promising methods for removing heavy metals from contaminated sediments (Chen and Lin, 2001).

This process uses the bacteria, principally *Thiobacillus ferrooxidans*, *Leptospirillum ferrooxidans* and thermophilic species of *Sulfobacillus*, *Acidianus* and *Sulfolobus*, to leach a metal of value such as copper, zinc, uranium, nickel and cobalt from a sulphide mineral. Bioleaching places the metal values of interest in the solution phase during oxidation. These solutions are handled for maximum metal recovery and the solid residue is discarded. It is a process of recovering of metals from a solid matrix into liquid, thus decontaminating the solid matrix, reducing the volume of the contaminated waste stream, recovery of the metal for reuse, reduces the risk of heavy metal exposures and bioavailability.

7. Landfarming

The biological removal of petroleum products using landfarming has been applied commercially in large scale with relative success. The technology has been widely used due to its simplicity and cost-effectiveness. Land farming is a bioremediation treatment process that is performed in the upper soil zone or in biotreatment cells. The dominant pollutant removal mechanisms involved in landfarming are volatilisation of low molecular weight volatile compounds during the early days of contamination or treatment, biodegradation and adsorption.

The petroleum products from the soil during landfarming are largely removed through volatilisation, biodegradation and adsorption (Morgan and Watkinson 1989; Devliegher and Verstraete 1996; Margesin et al. 1999). Lighter (more volatile) petroleum products like gasoline tend to be removed by volatilization during landfarm aeration process and to a lesser extent, degraded by microbial respiration (EPA, 1994). The mid-range petroleum products like diesel fuel and kerosene contain lower percentage of lighter constituents than does gasoline. Bio-degradation of these petroleum products is more significant than volatilization.

The more heavier or non volatile petroleum products like heating oil and lubricating oils do not volatilize during landfarm aeration, the dominant mechanisms that breaks down these petroleum products is biodegradation.

In this process contaminated soils are mixed with soil amendments such as soil bulking agents and nutrients, and then they are tilled into the earth. The material is periodically tilled for aeration.

Contaminants are degraded, transformed, and immobilized by microbiological processes and by oxidation. Soil conditions are controlled to optimize the rate of contaminant degradation. Moisture content, frequency of aeration, and pH are all conditions that may be controlled. Land Farming differs from composting because it actually incorporates contaminated soil into soil that is uncontaminated. Composting also generally takes place in aboveground piles.

Applications of Landfarming

This technique has been successfully used for years in the management and disposal of oily sludge and other petroleum refinery wastes. *In-situ* systems have been used to treat near surface soil contamination for hydrocarbons and pesticides.

The equipment employed in land farming is typical of that used in agricultural operations. These land farming activities cultivate and enhance microbial degradation of hazardous compounds. As a rule of thumb, the higher the molecular weight (i.e., the more rings within a polycyclic aromatic hydrocarbon), the slower the degradation rate. Also, the more chlorinated or nitrated the compound, the more difficult it is to degrade (Cookson, 1995). Bioremediation through landfarming is both simple and cost-effective to implement compared with other treatment technologies (Kelly et al. 1998; Pearce and Ollerman 1998).

On average, the costs associated with treating petroleum hydrocarbon-contaminated soil ranges from $30 to $70 per ton of contaminated soil compared with a physical treatment like soil venting which is relatively expensive ($70 to $200) per ton (Marijke and van Vlerken 1998).

However, as a result of costs associated with soil excavation and transporting the contaminated soil, in situ techniques can be in general about 40 – 50% of *ex-situ* techniques (SCG, 2004). The technology is simple in that typical equipments, which are used for landfarming, is used widely in the farming community and is therefore 'readily' available. As most of this equipment is designed to till the soil to a depth of 60.5 m, additional costs can be incurred during soil excavation for ex situ treatment (Kelly et al., 1998). For additional landfarm layouts or designs, the reader is referred to Doelman and Breedveld (1999) and to Battelle series (Alleman and Leeson 1999a–c). However, together with these advantages (Table 2), there are physical, chemical and biological aspects of the technology that can hamper the remediation process.

The physical aspects include the land area required for treatment, the ability and limitations of aeration equipments, mobility of pollutants in the soil, water requirements; chemical aspects include toxicity, transformation and partitioning of the petroleum products in different environmental media while biological aspects include biostimulation or bioaugumentation for optimal biotransformation of petroleum products in the soil.

Table 2. Benefits and constraints of landfarming

Benefits	Constraints
• Very low capital input required • Technology is simple to design and implement • Large soil volumes can be treated • Can be applied *ex-situ* • Has small environmental impact • Energy efficient	• Limited to removal of biodegradable pollutants • Large treatment area is needed • Involves risk of pollutant exposure • Substantial cost can be incurred during excavation • Limited knowledge of microbial process or the unraveling limiting factors during bioremediation

Limitations of Landfarming

The volatile organic compounds (VOCs) from the landfarm area can present air pollution problems if the treatment area is not properly covered to minimize the emissions. Apart from the VOC emissions, other constraints faced by the rehabilitation practitioners considering landfarming as a treatment option include, requirements for large land area for treatment, availability of the pollutant degrading bacteria, effectiveness of the technology at high constituent concentration (more than 50,000 ppm), improved concentration reductions in cases requiring more than 95% of pollution reduction and the flexibility of the technology in integrating the removal of petroleum hydrocarbons with other contaminants that may occur with the petroleum products. Although problems associated with depth of pollution can be solved by ex situ treatment, the polluted soil often requires a large treatment area, which can increase the risk of human exposure to the contaminants. However, such exposure is only temporary, as contaminants will be degraded if environmental conditions are optimal (Ausma et al., 2002).

Factors that may limit the applicability and effectiveness of the process include:

(1) Large space requirements;
(2) The conditions advantageous for biological degradation of contaminants are largely uncontrolled, which increases the length of time to complete remediation, particularly for recalcitrant compounds;
(3) Inorganic contaminants are not biodegraded;
(4) The potential of large amounts of particulate matter released by operations; and
(5) The presence of metal ions may be toxic to microbes and may leach from the contaminated soil into the ground.

8. Bioreactor

The term "bioreactor" in the context of soil and water bioremediation refers to any vessel or container where biological degradation of contaminants is isolated and controlled. Bioreactors can range from crude devices such as lined depressions in the ground to advanced metal containers where environmental conditions can be monitored and controlled. The essential treatment mechanism in a bioreactor is natural degradation by existing and/or added populations of microorganisms. Bioreactors have proven to be effective in remediating soil, and in some cases water, polluted with fuel hydrocarbons (oil, gasoline, and diesel) and organics. Slurry bioreactors are one of the most important types of *ad-situ* and *ex-situ* technique. Treatment of soils and sediments in slurry bioreactors (SBs) has become one of the best options for the bioremediation of soils polluted by recalcitrant pollutants under controlled environmental conditions (Mueller et al., 1991; Cookson, 1995, Mohan et al., 2006). SBs are also very often applied to determine the feasibility and actual potential of a biological strategy in the final restoration of a contaminated soil or site (Pinelli et al., 1997; Fava et al., 2004; 2005). In fact, under slurry conditions, the pollutant depletion rates depend mainly on the degradation activity of the microorganisms available in the system (Mueller et al., 1991) and the results obtained generally reflect the actual biological depuration potential of the soil (Bachmann et al., 1988; Evans et al., 1996; Tabak and Govind, 1997; Boopathy et al., 1997; Wenk et al., 1998).

However, SBs also have a few disadvantages, all of them related to requirements for soil excavation, handling, and conditioning, and bioreactor construction/operation that typically increase treatment costs compared to most simple bioremediation techniques (Mueller et al., 1991). In spite of this, SBs has resulted more cost effective than incineration, solvent extraction and thermal desorption in many cases (Castaldi, 2003).

9. Composting

Compost bioremediation refers to the use of a biological system of micro-organisms in a mature, cured compost to sequester or break down contaminants in water or soil. Micro-organisms consume contaminants in soils, ground and surface waters, and air. The contaminants are digested, metabolized, and transformed into humus and inert byproducts, such as carbon dioxide, water, and salts. Compost bioremediation has proven effective in degrading or altering many types of contaminants, such as chlorinated and nonchlorinated hydrocarbons, wood-preserving chemicals, solvents, heavy metals, pesticides, petroleum products, and explosives. Compost used in bioremediation is referred to as "tailored" or "designed" compost in that it is specially made to treat specific contaminants at specific sites.

Composting has been used as a bioremediation process for many different organic compounds. It is widely employed to recycle nutrients in garden and yard waste. Finished compost can be used as a soil conditioner. Extending composting technology to new bioremediation applications requires experiments. The biodegradation process must be effective within the context of existing environmental conditions, and odors and gases that are generated by the process have to be strictly controlled.

Composting also is considered as a remediation method for handling contaminated soil, sediment, and organic wastes (Williams and Keehan, 1993). Mechanical treatment by grinding, mixing, and sieving out nondegradable or disturbing materials (metals, plastics, glass, stones) gives good conditions for biological treatment of compostable materials (Zach, et al., 1999). The biological treatment builds up stable organic compounds through humification and reduces concentrations of organic pollutants. Composting can be used to lower the levels of chemical contaminants in residues or in soils to which polluted residues have been added.

So we can degrade the various pollutants by the help of micro organisms. Some examples are mention below.

Petroleum Biodegradation

Petroleum is a viscous liquid mixture that contains thousands of compounds mainly consisting of carbon and hydrogen. Petroleum and petroleum products constitute a variety of hydrocarbons, viz. xylanes, naphthalenes, octanes, camphor, etc. Petroleum-based products are the major source of energy for industry and daily life. Petroleum is also the raw material for many chemical products such as plastics, paints, and cosmetics. The transport of petroleum across the world is frequent, and the amounts of petroleum stocks in developed countries are enormous. Consequently, the potential for oil spills is significant, and research on the fate of petroleum in a marine environment is important to evaluate the environmental threat of oil spills, and to develop biotechnology to cope with them.

Crude oil is constituted from thousands of components which are separated into saturates, aromatics, resins and as phaltenes. Upon discharge into the sea, crude oil is subjected to weathering, the process caused by the combined effects of physical, chemical and biological modification. Saturates, especially those of smaller molecular weight, are readily biodegraded in marine environments. Aromatics with one, two or three aromatic rings are also efficiently biodegraded; however, those with four or more aromatic ring are quite resistant to biodegradation.

We can remediate the petroleum oil contaminated soil and water by the help of diverse group microorganism (Jain et al., 2011a). The microorganisms most responsible for bioremediation are bacteria. A large number of *Pseudomonas* strains capable of degrading PAHs have been isolated from soil and aquifers (Kiyohara et al., 1992: Johnson et al., 1996, Pathak et al., 2008). Other petroleum hydrocarbon degraders include *Yokenella* spp. *Alcaligenes* spp., *Roseomonas* spp., *Stenotrophomonas* spp., *Acinetobacter* spp., *Flavobacter* spp., *Corynebacterium* spp., *Streptococcus* spp., *Providencia* spp., *Sphingobacterium* spp., *Capnocy-tophaga* spp., *Moraxella* spp., *Bacillus* spp. and *Enterobacte*r spp. (Rusansky et al., 1987; Antai, 1990; Bhattacharya et al., 2002, Jain et al., 2010a,b). Other organisms such as fungi are also capable of degrading the hydrocarbons in engine oil to a certain extent. However, they take longer periods of time to grow as compared to their bacterial counterparts (Prenafeta-Boldu et al., 2001).

Petroleum Biodegradation in Marine Environments

Many catastrophic oil spills from large tanker accidents have attracted public attention to the fate of petroleum hydrocarbons in marine environments. Oil fields are not uniformly distributed around the globe, but being in limited areas such as the Persian Gulf region.

The world production of crude oil is more than three billion tons per year, and about the half of this is transported by sea. Consequently, the international transport of petroleum by tankers is frequent. All tankers take on ballast water which contaminates the marine environment when it is subsequently discharged.

More importantly, tanker accidents exemplified by that of the Exxon Valdez in Prince William Sound, Alaska, severely affect the local marine environment. Off-shore drilling is now common to explore new oil resources and this constitutes another source of petroleum pollution.

Table 3. Major genera of oil-degrading bacteria and fungi (Floodgate, 1984)

Bacteria	Fungi
Achrornobacter	Allescheria
Acinetobacter	Aspergillus
Actinomyces	Aureobasidium
Aeromonas	Botrytis
Alcaligenes	Candida
Arthrobacter	Cephalosporium
Bacillus	Cladosporium
Beneckea	Cunninghamella
Brevebacterium	Debaromyces
Coryneforms	Fusarium
Erwinia	Gonytrichum
Flavobacterium	Hansenula
Klebsiella	Helminthosporium
Lactobaoillus	Mucor
Leumthrix	Oidiodendrum
Moraxella	Paecylomyces
Nocardia	Phialophora
Peptococcus	Penicillium
Pseudomonas	Rhodosporidium
Sarcina	Rhodotorula
Spherotilus	Saccharomyces
Spirillum	Saccharomycopisis
Streptomyces	Scopulariopsis
Vibrio	Sporobolomyces
Xanthomyces	Torulopsis
	Trichoderma

However, the largest source of marine contamination by petroleum seems to be the runoff from land. Annually, more than two million tons of petroleum is estimated to end up in the sea. Fortunately, petroleum introduced to the sea seems to be degraded either biologically or abiotically (Readman et al., 1992).

In response to this concern, research into the biodegradation of petroleum in natural environments has been intensified.

The pioneering studies by Atlas, Bartha and their colleagues (Atlas and Bartha, 1993; Prince, 1993) have demonstrated that the available concentrations of nitrogen and phosphorus in seawater are limiting factors for the growth of hydrocarbon degrading microorganisms.

Thus, the addition of nitrogen and phosphorus fertilizers stimulates the biodegradation of petroleum. In general, small hydrocarbon molecules are more easily biodegraded than larger ones, and aromatics are degraded at a much slower rate than that of alkanes in marine environments (Oudot, 1984; Kennicut, 1988; Ishihara et al., 1995; Sugiura et al., 1997; Sasaki et al., 1998; Wang et al., 1998).

Many microorganisms capable of degrading petroleum components have been isolated. However, few of them seem to be important for petroleum biodegradation in natural environments. One group of bacteria belonging to the genus Alcanivorax does become predominant in an oil-contaminated marine environment, especially when nitrogen and phosphorus fertilizers are added to stimulate the growth of endogenous microorganisms. Hydrocarbon-degrading bacteria and fungi are widely distributed in marine habitats (Table 3) (Floodgate, 1984). However, their roles in a natural marine environment or in oil contaminated soil with or without treatment by bioremediation (fertilizer application to enhance microbial activities) are largely unknown.

Plastic Biodegradation

Biodegradation of plastics can be achieved by enabling microorganisms in the environment to metabolize the molecular structure of plastic films to produce an inert humus-like material that is less harmful to the environment. They may be composed of either bioplastics, which are plastics whose components are derived from renewable raw materials, or petroleum-based plastics. The use of bio-active compounds compounded with swelling agents ensures that, when combined with heat and moisture, they expand the plastic's molecular structure and allow the bio-active compounds to metabolize and neutralize the plastic. With more and more plastics being employed in human lives and increasing pressure being placed on capacities available for plastic waste disposal, the need for biodegradable plastics and biodegradation of plastic wastes has assumed increasing importance in the last few years. Biodegradable plastics are plastics that will decompose in natural aerobic (composting) and anaerobic (landfill) environments. Biodegradation of plastics can be achieved by enabling microorganisms in the environment to metabolize the molecular structure of plastic films to produce an inert humus-like material that is less harmful to the environment. They may be composed of either bioplastics, which are plastics whose components are derived from renewable raw materials, or petroleum-based plastics which utilize an additive. The use of bio-active compounds compounded with swelling agents ensures that, when combined with heat and moisture, they expand the plastic's molecular structure and allow the bio-active compounds to metabolize and neutralize the plastic.

Plastic can be degrade by *Streptomyces viridosporus, Streptomyces badius, Streptomyces setonii, Phanerochaete chrysosporium, Pelobacter venetianus* and *Bacteroides* etc. microorganism (Byungtae Lee et al., 1991)

Bioremediation of Phenol

Phenols are hydroxy compounds of aromatic hydrocarbons. Phenol derivatives are widely used as raw materials in many industries such as petrochemical, petroleum refineries, chemical and pharmaceutical. It is also widely used in pulp and paper mills, coking operations, coal refining, tannery and foundries. Phenolic compounds are among the most frequently found pollutants in rivers, industrial effluents, and landfill runoff waters. These compounds are toxic either by ingestion or by contact or inhalation even at low concentrations. Acute exposure of phenol causes central nervous system disorders, which leads to collapse and coma. Muscular convulsions with significant reduction in body temperature are also noted due to phenol toxicity, and this is known hypothermia. Renal damage and salivation may be induced by continuous exposure to phenol (Nair et al., 2008).

Phenols released into the environment are of high concern because of their potential toxicity (Boopathy, 1997). Degradation of phenol occurs as a result of the activity of a large number of microorganisms including bacteria, fungi and actinomycetes. (Bacterial species include *Bacillus* sp, *Pseudomonas* sp, *Acinetobacte*r sp, *Achromobacte*r sp etc. *Fusarium* sp, *Phanerocheate chrysosporium*, C*orious versicolo*r, *Ralstonia* sp, *Streptomyces* sp etc are also proved to be efficient fungal groups in phenol biodegradation. However, these microorganisms suffer from substrate inhibition at higher concentration of phenol, by which the growth is inhibited (Prieto et al., 2002). Many studies on biodegradation of phenol come from bacteria. The genus *Pseudomonas* is widely applied for the degradation of phenolic compounds. Microbiological degradation of phenol and some of its alkyl derivatives (p-cresol, 4-n-propyl phenol, 4-i –propyl phenol, 4-n-butyl phenol, 4-sec-butyl phenol, 4-t-butyl phenol and 4-t-octyl phenol) were examined under both aerobic and anaecrobic conditions in seven Japanese paddy soil samples (Atsushi et al., 2006). The rate of biodegradation of phenol by *Klebsiella oxytoca* strain was studied. It was found that *K. oxytoca* degraded phenol at elevated concentration where 75% of initial phenol concentration at 100 ppm was degraded within 72 h (Shawabkeh et al., 2007).

Enzyme Technology

Enzyme technology is a part of biotechnology that the European Federation of Biotechnology 1989 defined as-

"*Biotechnology is the integration of natural sciences and engineering sciences in order to achieve the application of organisms, cells, parts thereof and molecular analogues for products and services.*"

Microbes are still the most common source of industrial enzymes. Microorganisms produce enzymes inside their cells (intracellular enzymes) and may also secrete enzymes for action outside the cell (extracellular enzymes). The microorganisms selected are usually cultured in large fermentation chambers (known as fermenters) under controlled conditions to maximize enzyme production. The microorganisms may have specific genes introduced into their DNA through genetic engineering (Jain et al., 2011b), so that they produce specific pollutant degrading enzymes (Jain et al., 2010c). Enzymes are being used in numerous new applications in the food, feed, agriculture, paper, leather, and textiles industries, resulting in significant cost reductions. At the same time, rapid technological developments are now stimulating the chemistry and pharma industries to embrace enzyme technology, a trend strengthened by concerns regarding health, energy, raw materials, and the environment.

Enzymes as Decontaminating Agents

The use of enzymatic proteins may represent a good alternative for overcoming most disadvantages related to the use of microorganisms (Nannipieri and Bollag, 1991; Karam and Nicell, 1997; Nicell, 2001; Gianfreda and Bollag, 2002; Gianfreda and Rao, 2004). Enzymes have several beneficial characteristics (Table 4). They are the main effectors of all the transformations occurring in the biota. An enzyme has capability to produce extensive transformations of structural and toxicological properties of contaminants, and also they are able to their complete conversion into harmless inorganic end products. Enzymes can show advantages over traditional technologies and over microbial remediation also. Enzymes are resistance to inhibitors of microbial metabolism and can be used under extreme environmental conditions limiting microbial activity. They are effective at low pollutant concentrations and are active in the presence of microbial predators or antagonists. They act against a given substrate while microorganisms may prefer easily degradable compounds than the pollutant and are more mobile than microorganisms because of their smaller size (Gianfreda and Bollag, 2002).

Enzymes may act intracellularly, i.e. in the presence of or inside their originating cells; extracellularly, i.e. both in the presence or absence of their originating cells; free, i.e. soluble in solution and the catalysis will be homogenous; or immobilized, i.e. linked through different links to a solid matrix and the catalysis will be heterogeneous (Gianfreda and Rao, 2004).

Mostly hydrolases, dehalogenases, transferases and oxidoreductases are enzymatic classes in the remediation of polluted environments. Their main producers are bacteria, fungi, plants and microbe-plant associations. The breakdown of esteric, amidic and peptidic bonds by esterases, amidases and proteases may lead to products with little or no toxicity. Phosphotriesterases, amidases, proteases, carbohydrases (cellulases and amylases), depolymerase are belongs to the hydrolases. Mono- or di-oxygenases, reductases, dehalogenases, cytochrome P450 monoxygenases, phenoloxidases (laccases, tyrosinases) and peroxidases (lignin and manganese peroxidases) are oxidoreductases. The esteric, amidic and peptidic bond breakdowns by esterases, amidases and proteases respectively. Bacterial hydrolases such as carbamate or parathion hydrolases from *Achromobacter, Pseudomonas, Flavobacterium, Nocardia,* and *Bacillus cereus* have been successfully used in the transformation of pollutants such as carbofuran and carbaryl or parathion, diazinon and coumaphos (Coppella et al., 1990; Mulbry, and Eaton, 1991; Sutherland et al., 2002). Similarly, carbohydrases, depolymerases, proteases and phosphatases, produced by several bacteria and fungi, can be suitable for the transformation of insoluble materials such as carbohydrates, plastics and proteins (Nakamura et al., 2001; Singh, 2002). An interesting role is played by a class of enzymes involved in the transformation of nitrile compounds (Banerjee et al., 2002; Singh et al., 2006). Nitrile is any organic compound which has a -C≡N functional group. The -C≡N functional group. Many cyanide-containing compounds are highly toxic and deadly poisonous while some nitriles (which do not release cyanide ions) may have low toxicities. Nitrile compounds are synthesized by plants, fungi, bacteria, algae, insects and sponges. The prefix cyano is used in chemical nomenclature to indicate the presence of a nitrile group in a molecule.

Table 4. Application of oxidative enzymes (Duran and Esposito, 2000)

S. No.	Enzymes	Source	Applications
1.	Peroxidase	Horseradish *Artromyces ramosus* Plant materials	Degradation of phenols, chlorophenols, anilines. Decontamination of Kraft effluents and black liquor, biosensor phenol determination, dewatering of slimes. Degradation of phenols, polyaromatics and herbicides. Polymerization of humic acids Water decontamination
2.	Chloroperoxidase	*Caldariomyces funago*	Phenolic compounds oxidation. Biosensor chlorophenol detection
3.	Lignin peroxidase	*Phanerochaete chrysosporium* *Chrysonilia sitophila*	Degradation of aromatic compounds and phenolic materials. Remediation of Kraft effluent. Kraft effluent decontamination.
4.	Manganese peroxidase	*Phanerochaete chrysopsorium* *Nematolona frowardi e* *Phebia* *Radiate* *Lentinula edodes*	Degradation of phenols, lignin, pentachlorophenol. and dyes Lignin degradation Degradation of chlorophenol and diuron
5.	Tyrosinase	Sigma *Agaricus bisporus*	Phenol biosensor. Degradation of phenols and amines and removal of xenobiotic compounds. Oxidation of catechol and polymerization of phenolic compounds. Catechol oxidation.
6.	Laccase	*Trametes hispida* *Pyricularia oryzae* *Trametes versicolor* *Cerrena unicolor* *Pycnoporus cinnabarinus* Plant materials	Dyes decoloration. Azo-dyes degradation. Textile effluent degradation, chlorophenols and urea derivatives degradation. Phenol detoxification, 2,4-dichlorophenol degradation. Benzopyrenes degradation. Chlorophenols, xenobiotic binding to humus.
	Catechol dioxygenase	*Comamonas testosterone* *Pseudomonas pseudoalcaligenes*	Chlorophenol oxidation, diuron degradation. Polychlorinated biphenyls, chlorothanes.
7.	Phenoloxidaselike	*Gloeophyllum trabeum* *Trametes versicolor* *Phanerochaete chrysosporium* *Thermoascus aurantiacus*	Kraft effluent decontamination. Chlorinated compounds degradation. Kraft effluent.

Although extremely toxic, cyanide can enzymatically be converted to the less toxic thiocyanate by rhodaneses (thiosulfate:cyanide sulfurtransferases), highly conserved and widespread enzymes now considered as one of the mechanisms evolved for cyanide detoxification (Raybuck, 1992).

A genetic system was engineered to express high levels of recombinant *Pseudomonas aeruginosa* rhodanese (r- RhdA) in *E. coli*, and this organism was used to test the role of r-RhdA in cyanide detoxification (Cipollone et al., 2006).

Oxidative enzymes are mainly responsible for the detoxification of polluted environment. (Gianfreda and Bollag, 1994; Duran and Esposito, 2000; Torres et al., 1998; Gianfreda and Rao, 2004; Rodríguez Couto and Toca Herrera, 2006). These enzymes are involved in the formation of humus material in soil, in the exchanges between plants and soil through degradation and synthetic mechanisms. The main producers of oxidative enzymes of significance to the remediation of polluted environments are white-rot fungi (Reddy, 1995; Pointing, 2001; Asgher et al., 2008; Rubilar et al., 2008). White rot fungi are a genetically diverse group of wood degrading fungi that are found worldwide. Examples of white rot fungi include: *Phanerochaete chrysosporium, Ceriporiopsis subvermispora, Trametes versicolor* and *Pleurotus ostreatus*. Fungi, in particular of white rot basidiomycetes, *Phanerochaete chrysosporium* play an important environmental role in recycling wood and related materials, which is due primarily to the relatively non-specific processes used to initiate the degradation of the lignin fraction. The lignin-degrading or ligninolytic enzyme systems of *Phanerochaete chrysosporium*, determined by three enzymes-

(i) Lignin peroxidase (LiP)
(ii) Manganese dependent peroxidase (MnP)
(iii) Laccase (Lac)

(i) Lignin peroxidase (LiP, E.C. 1.11.1.14) - These are monomeric hemoproteins, found in many white rot fungi (Hammel and Cullen, 2008). These enzymes work when they are oxidized. Upon oxidation, lignin peroxidase releases free radicals capable of breaking apart other compounds which catalyze the one-electron oxidation of various lignin related substrates (Tien, 1987; Gold et al., 1989), then the resultant aryl cation radicals undergo spontaneous rearrangements and degradation. The ability of LiP to oxidize PAHs was first demonstrated by Hammel et al., . (1986), who observed transformation of those PAHs with ionization potential (IP) values of #7.55 eV. LiP is unable to degrade compounds above this threshold value; the metabolism of these compounds *in -vivo* therefore could not be explained by the direct action of this enzyme.

(ii) Manganese dependent peroxidase (MnP, E.C. 1.11.1.13) - The mechanism for the degradation of organic compounds by manganese peroxidase focuses on the manganese contained within the enzyme. The enzyme is capable of converting the manganese contained within the enzyme from Mn^{2+} to 6 Mn^{3+}. This Mn^{3+} can then oxidize lignin molecules. One electron oxidations of several polycyclic aromatics by Mn31 have been documented (Cavalieri and Rogan 1985; Cremonesi et al., 1989; Cremonesi et al., 1992), leading to the suggestion that the MnPs of *P. chrysosporium* might play a key role in the degradation of PAHs *in-vivo*. Manganese acetate, however, is finding to be incapable of oxidizing PAHs with IPs equal to or greater than that of chrysene (approximately 7.8 eV).

Thus, it is not clear that Mn31-mediated oxidations can explain the degradation of these PAHs (fluoranthene and phenanthrene) *in-vivo*. Recently, an *in-vitro* system mediated by the peroxidation of unsaturated lipids by MnP and Mn31 was described by Moen and Hammel (1994); this system catalyzed the oxidative cleavage of phenanthrene to diphenic acid, closely mimicking the action of intact fungal mycelium. These authors, therefore, invoked lipid peroxidation-based cooxidation to explain phenanthrene degradation by fungal cultures. The precise mechanism of the reaction, including the identity of the species which actually reacts with phenanthrene, has not yet been elucidated.

(iii) Laccase (Lac, E.C.1.10.3.2) - These are glycoproteins found throughout nature and are multicopper enzymes (Claus, 2004). The copper centers of these enzymes are able to produce highly reactive free radicals. These free radicals are capable of breaking apart covalent bonds found in many complex organic polymers, such as lignin and organic pollutants (Mayer and Staples 2002), The redox potential of laccase is rather low and less than those of nonphenolic substrates, which would be expected to limit their role in lignin degradation.

However, in the presence of an appropriate effective redox mediator, or laccase-mediator system, laccases can oxidize the non-phenolic lignin structure (Bourbannais and Paice, 1996). Due to their abilities to degrade lignin and other phenolic compounds, laccase has been widely applied to different biotechnological processes, such as biopulping with a patented Lignozyme® process and the degradation of synthetic dyes as part of the waste treatment process in the textile industry (Howard et al., 2003). Laccase is a polyphenol oxidase, but it is also nonspecific as to its reducing substrate and the variety of substrates oxidized includes polyphenols, methoxy-substituted phenols, diamines and a considerablerange of other compounds. It also acts as a mediator, enabling the oxidation of nonphenolic lignin model compounds that are not laccase substrates on their own (Thurston, 1994).

Above three enzymes are main components of these extracellular enzyme systems (Reddy, 1995; Koker et al., 2000). Moreover, the combined action of intracellular cytochrome P450 and extracellular laccase and peroxidase enzyme reactions enable white rot fungi to degrade a variety of toxic organopollutants, such as polychorinated phenols, polychlorinated dibenzo-p-dioxins, polycyclic aromatic hydrocarbon (PAHs) and other xenobiotics (Boominathan and Reddy, 1992). These enzymes are highly reactive and nonspecific free radicals, enabling fungi to degrade natural complex aromatic polymers of lignin. The polycyclic aromatic structure of both lignin and polycyclic aromatic hydrocarbons (PAHs) prompted the suggestion that the same fungi might be able to degrade these ubiquitous pollutants.

Therefore, *P. chrysosporium* is very interesting in applications for bioremediation and biotechnology (Koker et al., 2000). Under ligninolytic and nonligninolytic conditions, *P. chrysosporium* can metabolize a wide variety of PAHs, including the potent carcinogen benzo(a)pyrene (Haemmerli et al., 1986; Sanglard et al., 1986). Another impotant fungi involved in lignin biodegradation is *Pleurotus ostreatus*. The fungi, *P. ostreatus*, an edible, white rot fungus, commonly referred to as the oyster mushroom. It is useful in bioremediation because it can easily colonize contaminated soil and break down PAHs (Novotny et al., 1999; Eggen, 1999). *P. ostreatus* has been found to produce high levels of the extracellular enzyme laccase (Hou et al., 2004) and has the potential to completely mineralize PAHs (Bezalel et al., 1996). While screening for species able to degrade PAHs, Sack and Gunther (1993) found that *P. ostreatus* is quite efficient in the degradation of phenanthrene and fluorene and is less efficient with fluoranthene. Pyrene was not degraded to any significant extent. Vyas et al. (1994) showed the ability of *P. ostreatus* to degrade anthracene. These studies indicate the potential of *P. ostreatus* to degrade PAHs. *P. ostreatus* differs from *P. chrysosporium* in its lignin degradation mechanism in that it does not exhibit lignin peroxidase activity (Hatakka, 1994). Instead, its lignin degradation ability is assumed to be correlated with laccase activity (Higuchi, 1990; Thurston, 1994; Youn et al., 1995). For this reason, we can utilize the *P. ostreatus* to metabolize and mineralize PAHs. Since *P. ostreatus* is a noninvasive and non pathogenic fungus, commonly grown and eaten, it may be more readily accepted by the public for bioremediation applications.

Pleurotus ostreatus, Trametes versicolor and *Bjerkandera* sp. may be more promising than *P. chrysosporium* in their ability to mineralize PAHs to CO_2 (Field et al., 1992; Sack and Gunther, 1993; Vyas et al., 1994). In addition, *Crinipellis stipitaria* has been reported for its ability to metabolize pyrene (Lange et al., 1994; Lambert et al., 1994). Biodegradation of PAHs and PCBs also occurs with other white rot fungi, such as *Bjerkandara* sp. BOS55 (Kotterman et al., 1996), *Coriolus versicolor, Schizophyllum commune, Pycnoporus coccineus, Irpex lacteus, Bjerkandara adjusta, Coprinus cinereus, Ganoderma lucidum* (Matsubara et al., 2006), *Trametes versicolor, Coriolopsis polyzona* (Novotny et al., 1997) and *Phlebia brevispora* (Kamei et al., 2006).

Petroleum-Degrading Enzymes

Numerous microorganisms, such as bacteria, cyanobacteria, green algae, and fungi, are capable of degrading different components of petroleum under different environmental conditions (e.g., aerobic and anaerobic conditions at varied salinities and pHs). The enzymatic apparatus provides these capabilities to microorganisms. Petroleum degradation occurs gradually by sequential metabolism of its compounds.

The genes involved in degrading petroleum enzyme production may be located on chromosomal or plasmid DNA (Broderick, 1999). Biodegradation of hydrocarbons, both aliphatic and aromatic compounds, may occur under anaerobic or aerobic conditions (van Hamme et al., 2003). Under aerobic conditions, oxygenase enzymes introduce oxygen atoms into hydrocarbons (monooxygenases introduce one oxygen atom to a substrate while dioxygenases introduce two). The anaerobic degradation is catalysed by anaerobic bacteria, such as sulphate-reducing bacteria, using different terminal electron acceptors (van Hamme et al., 2003). The catechol di-oxygenase class of bacterial iron containing enzymes is an example of an enzyme class involved in the degradation of aerobic aromatic hydrocarbons. These enzymes are able to catalyse the addition of molecular oxygen atoms to 1, 2-dihydroxybenzene (catechol) and its derivatives, with subsequent cleavage of the aromatic ring (Madigan et al., 2010; Cao et al., 2009; van Hamme et al. 2003). Enzymes like catechol dioxygenases that are involved in aromatic ring cleavage are responsible for the wide variety of microorganisms capable of degrading aromatic compounds (Broderick, 1999).

Despite the fact that petroleum degradation under aerobic conditions occurs faster than under anaerobic conditions, it is important to note that anaerobic degradation is also essential to the bioremediation process because in several cases the environmental conditions can include limitations of the oxygen availability, such as in mangroves, aquifers and sludge digesters (Santos et al., 2011). In anaerobic metabolism, generally, aromatic compounds are converted into benzoyl-CoA, which is target of the benzoyl-CoA reductase (BCR) action (Hosoda et al., 2005). As an example of enzymatic bioremediation, PAH detoxification can be achieved by the use of laccases (Alcalde et al., 2005) (enzymes capable of catalysing the oxidation of phenols, polyphenols, and anilines, coupled to the 4-electron reduction of molecular oxygen to water) (Smith et al., 1997). A great advantage of the enzymatic bioremediation of xenobiotics that are either hydrophobic or poorly soluble in aqueous solutions, such as PAHs, is that enzymatic oxidation can occur in the presence of organic solvents (Alcalde et al., 2005).

Plastic Degrading Enzymes

Polyurethane (PUR), which is a kind of plastic, is widely used as base material in various industries. PUR is synthesized by condensation of polyol and polyisocyanate. polyester polyurethane (PUR)-degrading enzyme, PUR esterase, derived from *Comamonas acidovorans* TB-35, a bacterium that utilizes polyester PUR as the sole carbon source, was purified (Yukie, 1997). There are two types of PUR, named according to the polyol used in the synthesis: PUR which uses polyester as polyol in synthesis is called polyester PUR, and the one which uses polyether is called polyether PUR. PUR is a material which is resistant to microbial attack, but generally, the ester-type PUR is more easily degraded than the ether-type PUR (Morton, 1994). Several kinds of fungi and bacteria that can degrade ester-type PUR have been reported (Darby and Kaplan, 1968; Pathirana and Seal, 1984; Kay et al., 1991; Crabbe et al., 1994). The degradation was considered to be initiated by hydrolysis of the ester bond by some hydrolytic enzyme(s), such as esterase. Degradation of poly(3-hydroxyalkanoate) (PHA) by PHA depolymerase is well-known (Tanio et al., 1982; Mukai et al., 1992; Muller and Jendrossek, 1993; Schirmer et al., 1993; Kanesawa et al., 1994; Kasuya and Doi, 1994; Kita et al., 1995; Nojima et al., 1996). This enzyme, which is one of the esterases, has a hydrophobic surface-binding domain and a catalytic domain and accomplishes the degradation of PHA film via two steps (Mukai et al., 1993; Kasuya et al., 1995; 1996). The first step is adsorption of the enzyme on the surface of the film via the surface-binding domain, and the second step is hydrolysis of the ester bond. Activities of several enzymes, namely, esterase, protease, and urease in culture broth during ester-type PUR degradation by fungi is also reported by many observers. In another study degradability of four kinds of commercial biodegradable plastics (BPs), polyhydroxybutyrate and hydroxyvalerate (PHBV) plastic, polycaprolactone plastic (PCL), blend of starch and polyvinyl alcohol (SPVA) plastic and cellulose acetate (CA) plastic were investigated, PHBV was reported to be easily degradable in soil or under composting conditions (Margaret et al., 1993; Yue et al., 1996). Some studies also showed the complete degradation of PHBV in 28 days by a conventional degradability assay using activated sludge (Sugano et al., 1999). Ishigaki et al. (1999, 2002) reported that augmentation of effective microbe or enzyme rather than stimulation seemed necessary to achieving the fragmentation of plastic using natural polymer such as SPVA and CA.

Biodegradation of Phenol

Bioremediation of phenol occurs as a result of the activity of a large number of microorganisms including bacteria, fungi and actinomycetes Bacterial species include *Bacillus* sp., *Pseudomonas* sp., *Acinetobacte*r sp., *Achromobacte*r sp. , *Fusarium* sp., *Phanerocheate chrysosporium*, C*orious versicolo*r, *Ralstonia* sp., *Streptomyces* sp. etc are also proved to be efficient fungal groups in phenol biodegradation.

Table 5. Enzymes involved in the biodegradation of phenol

S. No.	Enzymes	References
1.	Phenol hydroxylase	Gurujeyalakshmi and Oriel (1988)
2.	Polyphenol Oxidase	Burton et al. (1993)
3.	Polyphenol Oxidase	Cano et al. (1997)
4.	Phenol Oxidase	Okeke et al. (1997)
5.	Polyphenol oxidase	Shashirekha et al. (1997)
6.	Catechol 2,3 dioxygenase	Ali et al. (1998)
7.	Laccase	Bollag et al. (1998)
8.	Polyphenol oxidase	Garzillo et al. (1998)
9.	Peroxidase	Ghioureliotis and Nicell (1999)
10.	Horse radish peroxidase	Wu et al. (1998)
11.	Horse radish peroxidase	Zahida et al. (1998)
12.	Polyphenol oxidase	Edwards et al. (1999)
13.	Laccase	Kadhim et al. (1999)
14.	Laccase	Schneider et al. (1999)
15.	Laccase	Hublik and Schinner (2000)
16.	Laccase	Robles et al. (2000)
17.	Catechol 1,2oxygenase	An et al. (2001)
18.	Polyphenol oxidase	Luke and Burton (2001)
19.	Polyphenol oxidase	Steffens (2002)
20.	Phenol oxidase	Johjima et al. (2003)
21.	Tyrosinase	Xiangchun (2003)

However, these microorganisms suffer from substrate inhibition at higher concentration of phenol, by which the growth is inhibited (Prieto et al., 2002). These microorganisms are capable of degrading phenol through the action of variety of enzymes. These enzymes may include oxygenases hydroxylases, peroxidases, tyrosinases and oxidases (Table 5). Oxygenases include monoxygenases and dioxygenases. Most of the aromatic catabolic pathways converge at catechol. Catechols are formed as intermediates from a vast range of substituted and nonsubstituted mono and poly aromatic compounds. Aerobically, phenol also is first converted to catechol, and subsequently, the catechol is degraded via ortho or meta fission to intermediates of central metabolism. The initial ring fission is catalysed by an ortho cleaving enzyme, catechol 1, 2 dioxygenase or by a meta cleaving enzyme catechol 2,3 dioxygenase, where the product of ring fission is a cis-muconic acid for the former and 2-hydro cis muconic semi aldehyde for the latter (Gurujeyalakshmi and Oriel, 1988). *Streptomyces setonii* (ATCC 39116) degraded aromatic compounds such as phenol or benzoate via an ortho cleavage pathway using catechol 1, 2 dioxygenase (An et al., 2001).

Phenol hydroxylase (E. C 14. 1.3.7) catalyses the degradation of phenol via two different pathways initiated either by ortho or meta cleavage.

Problems Associated with the Use of Enzymes

Despite all the advantages related to enzymatic bioremediation, high production costs, low yields, and enzymatic inhibition are some of the problems that must be overcome. The use of an enzyme is possible if the product of the enzyme-mediated reaction is less toxic than the substrate. Detoxification requires a multistep process, i.e. many enzymes acting sequentially, only specific microorganisms are suitable for achieving decontamination. Even if enzymes require cofactors, their use may be problematic, unless a preparation containing both the enzyme and the respective cofactor is used. Another problem in the use of enzymes to detoxify organic-polluted soil is given by the rapid degradation of the free enzyme by proteases released by soil microorganisms. In natural environments, disadvantages that may hinder or diminish the catalytic potential of enzymatic catalysts may depend on both the pollutants to be transformed and the enzymes. In a polluted site, mixtures or composed combinations of many organic(and inorganic) contaminants, rather than a single pollutant, are present and the complexity of the pollution may entail possible negative or positive, synergistic effects on the enzyme efficiency. Enzymes may reduce or even lose their activity upon pollutant transformation or they may present a low stability and survival under very often harsh environmental conditions. If their repeated use is required, enzymes may present low reusability, thus decreasing the efficiency of the whole treatment. In addition, if isolated enzymes are used, the cost of enzyme isolation and purification greatly hampers their practical application, mainly when their continuous feeding is needed. From an environmental point of view, the use of enzymes instead of chemicals or microorganisms undoubtedly presents some advantages (Ahuja, 2004):

(i) The biotransformation does not generate toxic sideproducts as is often the case with chemical and some microbiological processes, and the enzymes are digested, in situ, by the indigenous microorganisms after the treatment.

(ii) The requirement to enhance bio-availability by the introduction of organic co-solvents or surfactants is much more feasible from an enzymatic point of view than using whole cells.

(iii) The potential to produce enzymes at a higher scale, with enhanced stability and/or activity and at a lower cost by using recombinant-DNA technology.

Genetic Engineering Approaches

During the past 20 years, recombinant DNA techniques have been studied intensively to improve the degradation of hazardous wastes under laboratory conditions. Only one field test has been successfully implemented. Recombinant bacteria can be obtained by genetic engineering techniques or by natural genetic exchange between bacteria. Genetically engineered microorganisms (GEMs) have shown potential for bioremediation applications in soil, groundwater, and activated sludge environments, exhibiting enhanced degradative capabilities encompassing a wide range of chemical contaminants.

However, the vast majority of studies pertaining to genetically engineered microbial bioremediation are supported by laboratory-based experimental data. In general, relatively few examples of GEM applications in environmental ecosystems exist. Unfortunately, the only manner in which to fully address the competence of GEMs in bioremediation efforts is through long-term field release studies. It is therefore essential that field studies be performed to acquire the requisite information for determining the overall effectiveness and risks associated with GEM introduction into natural ecosystems. The use of genetic engineering to create organisms specifically designed for bioremediation has great potential. The bacterium *Deinococcus radiodurans* (the most radioresistant organism known) has been modified to consume and digest toluene and ionic mercury from highly radioactive nuclear waste. Most commonly, the process is misunderstood. The microbes are ever-present in any given context and are generally referred to as "normal microbial flora". During bioremediation (biodegradation) processes, fertilizer/nutrient supplementation is introduced to the environments in efforts to maximize growth and production potential. Common misbelief is that microbes are transported and dispersed into an unadulterated environment.

Enzymatic bioremediation improved with molecular tools can be particularly suitable for situations where rapid remediation is required (Sutherland et al., 2004). Alcalde and colleagues (Alcalde et al., 2006) reported that recent studies of protein engineering, metagenomics, and proteomics are effectively contributing to cost reduction, minimizing chemical use and also improving costbenefit ratios. The use of molecular tools for biocatalysis applications can also help solving the problem of GMO use in the environment (Alcalde et al., 2006); for instance, if the production of a modified enzyme is performed *in vitro*, it is not necessary to introduce the modified organism into the natural environment.

There are at least four principal approaches to GEM development for bioremediation application. These include:

(1) Modification of enzyme specificity and affinity,
(2) Pathway construction and regulation,
(3) Bioprocess development, monitoring, and control,
(4) Bioaffinity bioreporter sensor applications for chemical sensing, toxicity reduction and end point analysis.

Conclusion

Bioremediation remains a field with much work to be done, with few extremely effective field applications due to the extremely diverse conditions found in different ecosystems, the development of alternative or complementary strategies is continually encouraged. Bacteria play a central role in the biogeochemical cycling of metals in the environment, yet the molecular and biochemical mechanisms for most of these processes are not well understood. We should recognize that, in the past few years, enzymatic bioremediation has become an attractive alternative to further support the bio-treatment techniques currently available – enzymes provide simpler systems than a whole organism (Pieper, 2004). Most xenobiotics can be submitted to enzymatic bioremediation, for example, polycyclic aromatic hydrocarbons (PAHs) (Samanta, 2002), polynitrated aromatic compounds (Ramos, 2005), pesticides such as organochlorine insecticides (Ahuja, 2004), bleach-plant effluents, synthetic dyes, polymers and wood preservatives (creosote, pentachlorophenol) (Bajpai, 2004).

The benefits provided by molecular tools can open unlimited windows of opportunity, as it is possible to detect genes from cultivable or noncultivable organisms (using metagenomics) and to express these genes in cultivable organisms, using enzymes that were not yet described.

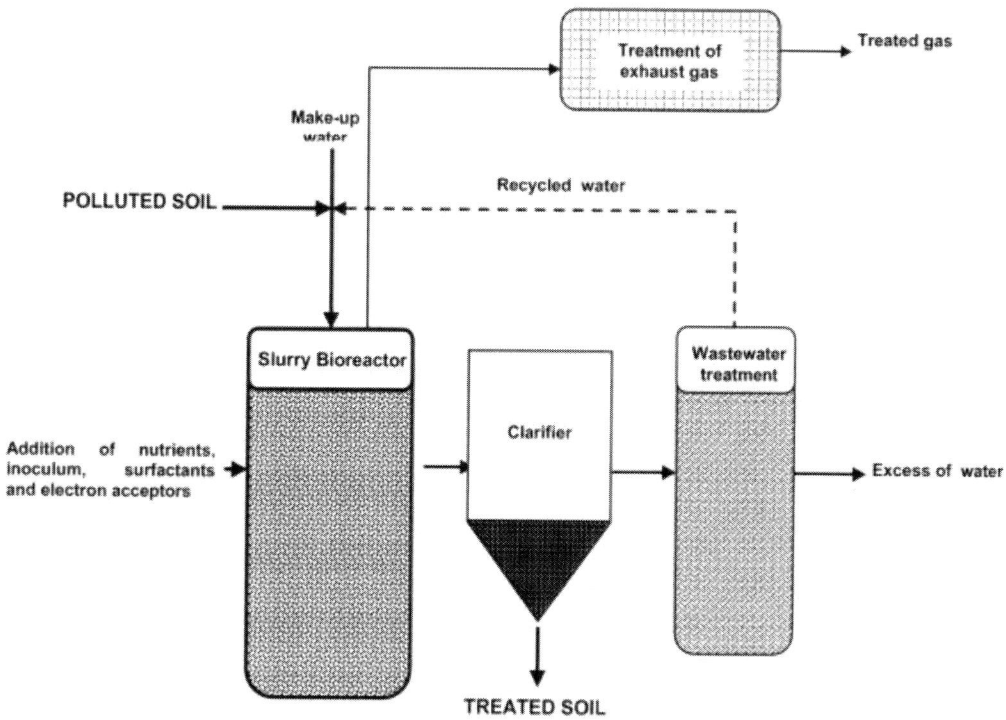

Figure 1. Slurry Bioreactors (SBs).

Enzymatic bioremediation also influences other biological areas, such as medicine, since Rittmann and Schloendorn (Rittmann and Schloendorn, 2007) proposed the idea of "medical bioremediation," based on and inspired by environmental bioremediation principles. Several external factors can limit the rate of biodegradation of organic compounds. These factors may include temperature, pH, oxygen content and availability, substrate concentration and physical properties of contaminants. Each of these factors should be optimized for the selected organism for the maximum degradation of the organic compound of choice. The optimization of the substrate concentration in phenol biodegradation is particularly important since it inhibits the growth of the organism at higher concentrations.

References

Ahuja, S.K. 2004. Utilization of enzymes for environmental applications. *Crit. Rev. Biotechnol.,* 24, 125–154.

Alcalde, M., Ferrer, M., Plou, F. J. and Ballesteros, A. 2006. Environmental biocatalysis: from remediation with enzymes to novel green processes, *Trends in Biotechnology,* 24 (6), 281–287.

Alcalde, M.T., Bulter, M. and Zum-arraga. 2005. Screening mutant libraries of fungal laccases in the presence of organic solvents. *Journal of Biomolecular Screening,* 10 (6), 624–631.

Alexander, M. 1965. Biodegradation: Problems of molecular recalcitrance and microbial infallibility. *Adv. Appl. Microb.,* 7, 35-80.

Alfoldi, L. 1988. Groundwater microbiology: problems and biological treatment-state-of-the-art report. *Water Sci. Technol.,* 20, 1–31.

Ali, S., Roberto, F., La fuente and Dona, A.C. 1998. Meta pathway degradation of phenolics by thermophilic *Bacilli. Enzyme Microbial Technol.,* 23, 462-468.

Alleman, B.C.. and Leeson, A. 1999a. Bioreactor and ex-situ biological treatment technologies. The fifth International in-situ and on site bioremediation symposium, San Diego, California, USA, 19–22. Battelle Press.

Alleman, B.C.. and Leeson, A. 1999b. Bioremediation technologies for polycyclic aromatic hydrocarbons compounds. The fifth International insitu and on site bioremediation symposium, San Diego, California, USA, 19–22. Battelle Press.

Alleman, B.C.. and Leeson, A. 1999c. In-situ bioremediation of petroleum hydrocarbons and other organic compounds. The fifth International insitu and on site bioremediation symposium, San Diego, California, USA, April 19–22. Battelle Press.

An, H., Park, H. and Kim, E., 2001. Cloning and expression of thermophilic catechol 1,2 dioxygenase gene (cat A) from *Streptomyces setonii. FEMS Microbiol. Lett.,* 195, 17-22.

Antai, S.P. 1990. Biodegradation of Bonny light crude oil by *Bacillus* sp. and *Pseudomonas* spp. *Waste Manag.,* 10: 61-64.

Asgher, M., Nawaz-Bhatti, H., Ashraf, M. and Legge, R. L. 2008. Recent developments in biodegradation of industrial pollutants by white rot fungi and their enzyme system. *Biodegradation,* 19, 771-783.

Atlas, R.M.. and Bartha, R. 1993. Microbial Ecology: Fundamentals and Applications. 3rd ed. *Addison Wesley Longman Publishers*. Amsterdam.

Atsushi, S., Yasushi, I. and Arata, K. 2006. Aerobic and anaerobic bio degradation of phenol derivatives in various paddy soils. *Sci. total Environ.,* 367, 979-987.

Ausma, S., Edwards, G.C., Fitzgerald-Hubble, C.R, Halfpenny-Mitchell, L., Gillespie, T.J.. and Mortimer, W.P. 2002. Volatile hydrocarbon emissions from a diesel fuel contaminated soil bioremediation facility. J. *Air Waste Manage. Assoc.,* 52, 769–780.

Bachmann, A., De-Bruin, W., Jumelet, J.C., Rijnaarts, H.H. and Zehnder, A.J. 1988**.** Aerobic biomineralization of alpha-hexachlorocyclohexane in contaminated soil. *Appl. Environ. Microbiol.,* 54(2)**,**548-554.

Bajpai, P. 2004. Biological bleaching of chemical pulps. Crit. Rev. *Biotechnol.,* 24, 1–58.

Bajpai, V., Sharma, Pallavi, Jain, Pankaj-Kumar and Gaur Rajarshi K. 2011. Bioremediation of heavy metals - microbial application. In Recent Advances in Environmental

Biotechnology, Edited by Jain, P. K.,V. K. Gupta and V Bajpai. pp. 205-218. LAP Lambert Academic Publishing AG and Co. KG, Germany.

Banerjee, A., Sharma, R. and Banerjee, U.C. 2002. The nitrile-degrading enzymes: current status and future prospects. *Appl. Microbiol. Biotechnol.,* 60, 33-44.

Bezalel, L., Hadar, Y. and Cerniglia, C.E. 1996. Mineralization of Polycyclic Aromatic Hydrocarbons by the White Rot Fungus *Pleurotus ostreatus. Applied and Environmental Microbiology,* 62, 292-295.

Bhattacharya, D., Sarma, P.M., Krishnan, S., Mishra, S. and Lal, B. 2002. Evaluation of genetic diversity among *Pseudomonas citronellolis* strains isolated from oily sludge contaminated sites. *Applied Environ. Microbiol.,* 663: 1435-1441.

Bioremediation of Chlorinated Solvents Consortium of the Remediation Technologies. 1996. Development Forum (RTDF). DRAFT Guidance Handbook on Natural Attenuation of Chlorinated Solvents.

Bollag, J.M., Shuttle, W.K.N.. anderson. D.H. 1998. Laccase mediated detoxification of phenolic contaminants. Appl. *Environ. Microbiol.,* 54(12), 3086-3091.

Boominathan, K.. and Reddy, C.A. 1992. Fungal degradation of lignin: biotechnological applications. Arora, D.K., Elander, R.P., Mukerji, K.G. (Eds), *In handbook of applied mycology,* 4, 763-822. Marcel Dekker Inc., New York.

Boopathy, R. 1997. Anaerobic Phenol Degradation by Microorganisms of Swine Manure. *Current Microbiology,* 35, 64-6.

Boopathy, R., Manning , J. and Kulpa, C.F. 1997.Optimization of environmental factors for the biological treatment of trinitrotoluene-contaminated soil. *Arch. Environ. Contam. Toxicol.,* 32(1), 94-98.

Bourbannais, R.. and Paice, M.G. 1996. Enzymatic delignification of Kraft pulp using laccase and a mediator. *TAPPI. J.,* 79, 199-204.

Broderick, J.B. 1999. Catechol dioxygenases, *Essays in Biochemistry,* 34 (11), pp. 173–189.

Burton, S.G., John, R.D., Perry, T.K. and Peter, D.R. 1993. Activity of mushroom polyphenol oxidase in organic medium. *Biotechnol. Bioeng.,* 42, 938-944.

Byungtae-Lee, Anthony, L., Pometto, I.I.I., Alfred-Fratzke. and Theodore, B., Bailey, Jr. 1991. Biodegradation of Degradable Plastic Polyethylene by *Phanerochaete* and *Streptomyces*Species, *Appl. Environ. Microbiol.,* 57(3), 678–685.

Cano, P.M., Begona-de-Ancos, Gloria, L.M. and Mariana, S. 1997. Improvement of Frozen banana (*Musa cavendisher, C. Venana*) colour by branching, relationship between browning, phenols and poly phenol oxidase and peroxidase activities. *Z Lebnsm unters Forsch A,* 60-65.

Cao, B., Nagarajan, K.. and Loh, K.C. 2009 Biodegradation of aromatic compounds: current status and opportunities for biomolecular approaches. *Applied Microbiology and Biotechnology,* 85(2), 207–228.

Castaldi, F.J. 2003. Tank-based bioremediation of petroleum waste sludges. *Environ Prog.,* 22(1), 25-36.

Castaldi, F.J. and Ford, D.L. 1992. Slurry bioremediation of petrochemical waste sludges. *Water Sci Technol.,* 25(3), 207-212.

Cavalieri, E.L.. and Rogan, E.G. 1985. Role of radical cations in aromatic hydrocarbon carcinogenesis. *Environ. Health Perspect.,* 64, 69–84.

Chen, S.Y. and Lin,J.G. 2001. Bioleaching of heavy metals from sediment: significance of pH. *Chemosphere,* 44, 1093-1102.

Cipollone, R., Ascenzi, P., Frangipani, E. and Visca, P. 2006. Cyanide detoxification by recombinant bacterial rhodanese. *Chemosphere,* 63, 942-949.

Claus, H. 2004. Laccases: structure, reactions and distribution. *Micron.,* 35, 93-96.

Cookson, J.T.1995. Bioremediation Engineering: Design and Application McGraw-Hill, New York, USA.

Coppella, S.J., Cruz, N.D., Payne, G.F., Pogell, B.M., Speedie, M.K. Karns, J.S., Sybert, E.M. and Connor, M.A. 1990. Genetic Engineering approach to toxic waste management case study for organophosphate waste treatment. *Biotechnol. Prog.,* 6, 76-81.

Crabbe, J.R., Campbell, Thompson, L., Walz, S.L. and Schultz, W.W. 1994. Biodegradation of a colloidal ester-based polyurethane by soil fungi. Int. *Biodeterior. Biodegrad.,* 33,103–113.

Crawford, R.L. and Crawford, D.L. 1996.*Bioremediation*: Principles and Applications Cambridge University Press. United Kongdom.

Cremonesi, P., Cavalieri, E.L. and Rogan, E.G. 1989. One-electron oxidation of 6- substituted benzo(*a*)pyrenes by manganic acetate. A model for metabolic activation. *J. Org. Che.,* 54, 3561–3570.

Cremonesi, P., Heitbrink, B., Rogan, E.G. and Cavalieri, E.L. 1992. One electron oxidation of dibenzo(*a*)pyrenes by manganic acetate. *J. Org. Chem.,* 57, 3309–3312.

Darby, R. T.. and Kaplan, A.M. 1968. Fungal susceptibility of polyurethanes. Appl. *Microbiol.,* 16, 900–905.

DeVliegher, W. and Verstraete, W. 1996. Formation of non-bioavailable organic residue in soil: *Perspectives for site remediation. Biodegradation,* 7, 471–485.

Dhaker, A.K. and Jain, P.K. 2011. Sewage pollutants and their bioremediation by using different microorganism. In Recent Advances in Environmental Biotechnology, Edited by Jain, P.K., V.K. Gupta.. and Bajpai, V., pp. 189-204. LAP Lambert Academic Publishing AG and Co. KG, Germany.

Doelman, P. and Breedveld, G. 1999. In situ versus on site practices. Bioremediation of contaminated soils. Agronomy monograph no. 372. pp. 539-558. Soil Science Society of America, Madison, USA.

Duran, N. and Esposito, E. 2000. Potential applications of oxidative enzymes and phenoloxidase-like compounds in wastewater and soil treatment: a review. *Appl. Catal. B. Enzym.,* 28, 83-99.

Edwards, W., Bowness, R., Leukes, W.D., Jacobs, E.P., Sanderson, R., Rose, P.D. and Burton, S.G. 1999a. A capillary membrane reactor using immobilized polyphenol oxidase for the removal of phenols from industrial effluents. *Enzyme Microbial Technol.,* 24, 209-217.

Edwards, W., Lewkes, W.D., Rose, P.D. and Burton, S.G. 1999b. Immobilization of polyphenol oxidase on chitosan coated polysulphone capillary membrane for improved phenolic effluent bioremediation. *Enzyme Microbial Technol.,* 25, 769-773.

Eggen, T. 1999. Application of fungal substrate from commercial mushroom production – Pleurotus ostreatus –for bioremediation of creosote contaminated sediment. *International Biodeterioration and Biodegradation,* 44, 117-126.

Environmental Protection Agency (EPA). 1994. Landfarming In: How to Evaluate Alternative Cleanup Technologies for Underground Storage Tank Sites: A guide for Corrective Action Plan reviews (EPA 510-B-94-003 and EPA 510-B-95-007).

Evans, B.S., Dudley, C.A. and Klasson, K.T.1996. Sequential anaerobic-aerobic biodegradation of PCBs in soil slurry microcosms. *Appl. Biochem Biotechnol*., **57–**58,885-894.

Eweis, J.B., Ergas, S.J., Chang, D.P.Y. and Schroeder, E.D.1998. Bioremediation Principles McGraw-Hill. USA, 207-236.

Fava, F. and Di-Gioia, D. 2005. Marchetti L: Role of the reactor configuration in the biological detoxification of a dump site-polychlorobiphenyl-contaminated soil in lab slurry phase conditions. *Appl Microbiol Biotechnol*., 53(2),243-248.

Fava, F., Berselli, S., Conte, P. and Piccolo, A. 2004. Marchetti, L. Effects of humic substances and soya lecithin on the aerobic bioremediation of a soil historically contaminated by polycyclic aromatic hydrocarbons (PAHs). *Biotechnol Bioeng*., 88(2), 214-223.

Fewson, C.A. 1988. Biodegradation of xenobiotic and other persistent compounds: the causes of recalcitrance. *Trends Biotechnol*., l6, 148–153.

Field, J.A., Jong, de, E., Costa, G.F. and Bont, J.A.M. de. 1992. Biodegradation of polycyclic aromatic hydrocarbons by new isolates of white rot fungi. *Appl. Environ. Microbiol*., 58, 2219–2226 .

Floodgate, G.D. 1984. The Fate of Petroleum in Marine EcosystemshnPetroleum Microbiology, R.M. Atlas (ed.), pp 373, New York, NY: Macmillan Publishing Co.

Fuller, M.E. and Manning, J.F. 2004. Microbiological changes during bioremediation of explosives-contaminated soils in laboratory and pilot-scale bioslurry reactors. *Bioresour. Technol*., 91(2), 123-133.

Garzillo, A.M.V, Colao, M.C, Caruso, C., Caporate, C., Celletti, D. and Buonocore, V. 1998. Laccase from the white rot fungus *Trametes trogii*. *Appl. Microbiol. Biotechnol*., 49, 545-551.

Ghioureliotis, M. and Nicell, J. 1999. Assessment of soluble products of peroxidase catalysed polymerisation of aqueous phenol. *Enzyme Microbial Technol*., 25, 185-193.

Gianfreda, L. and Bollag, J.M. 1994. Effect of soils on the behavior of immobilized enzymes. *Soil Sci. Soc. Am. J*., 58, 1672-1681.

Gianfreda, L. and Bollag, J.M. 2002. Isolated enzymes for the transformation and detoxification of organic pollutants. In: R.G. Burns, R. Dick (eds). Enzymes in the Environment: Activity, Ecology and Applications. pp 491-538, Marcel Dekker, New York.

Gianfreda, L. and Rao, M.A. 2004. Potential of extra cellular enzymes in remediation of polluted soils: a review. *Enzyme Microb. Technol*., 35, 339-354.

Gold, M.H., Wariishi, H. and Valli, K. 1989 Extracellular peroxidases involved in lignin degradation by the white rot basidiomycete *Phanerochaete chrysosporium*. *ACS Symp. Ser*., 389, 127-140.

Gurujeyalekshmi, G. and Oreil, P. 1988. Isolation of phenol degrading Bacillus stearothermophilus and partial characterization of the phenol hydroxylase. Appl. *Environ. Microbiol*., 55(2), 500-502.

Haemmerli, S.D., Leisola, M.S.A., Sanglard, D. and Fiechter A. 1986. Oxidation of benzo(a)pyrene by extracellular ligninase of Phanerochaete chrysosporium: veratryl alcohol and stability of ligninase. *J. Biol. Chem*., 261, 6900–6903.

Hammel, K.E. and Cullen, D. 2008 . Role of fungal peroxidases in biological ligninolysis. *Current Opinion in Plant Biology, 11*, 349-355.

Hammel, K.E., Kalyanaraman, B. and Kirk, T.K. 1986. Oxidation of polycyclic aromatic hydrocarbons and dibenzo(*p*)dioxins by *Phanerochaete chrysosporium* ligninase. *J. Biol. Chem.*, 261, 16948–16952.

Hatakka, A., 1994. Lignin-modifying enzymes from selected white-rot fungi: production and role in lignin degradation. *FEMS Microbiol. Rev., 13,* 125– 135.

Higuchi, T. 1990. Lignin biochemistry: biosynthesis and biodegradation. *Wood Sci. Technol.,* 24, 23–63.

Hinchman, R. and Negri, C. 1994. The Grass Can Be Cleaner on the Other Side of the Fence. 1994. *logos, Argonne National Laboratory,* 12(2), 8-11.

Hosoda, A., Kasai, Y., Hamamura, N., Takahata, Y. and Watanabe, K. 2005. Development of a PCR method for the detection and quantification of benzoyl-CoA reductase genes and its application to monitored natural attenuation. *Biodegradation,* 16(6), 591–601.

Hou, H., Zhou J., Wang J., Du C. and Yan, B. 2004. Enhancement of laccase production by *Pleurotus ostreatus* and its use for the decolonization of anthraquinone dye. *Process Biochemistry,* 39, 1415-1419.

Howard, R.L., Abotsi, E., Jansen, van., Rensburg, E.L. and Howard, S. 2003. Lignocellulose biotechnology: issues of bioconversion and enzyme production. *Afr. J. Biotechnol.,* 2, 602-619.

Hublik, G. and Schinner, F. 2000. Characterization and immobilization of the laccase from *Pleurotus ostreatus* and its use for the continuous elimination of phenolic pollutants. *Enzyme Microbial. Technol.,* 27, 330-336.

Ishigaki, T., Kawagoshi, Y. Ike, M. and Fujita, M., 1999. Biodegradation of polyvinyl alcohol–starch blend plastic film. *World J. Microbiol. Biotechnol.,* 15, 251–257.

Ishigaki, T., Sugano, W., Ike, M. and Fujita, M. 2002. Effect of UV irradiation on enzymatic degradation of cellulose acetate. *Polym. Degrad. Stab.*, 78, 505–510.

Ishihara, M., Sugiura, K., Asaumi, M., Goto, M., Sasaki, E. and Harayama, S. 1995. Oil degradation in microcosms and mesocosms. *In: Microbial Processes for Bioremediation,* 15, 101–116.

Jain, P.K., Gupta, V.K., Bajpai, V., Lowry, M. and Jaroli, D.P., 2011b. GMO's: In perspective of bioremediation. In Recent Advances in Environmental Biotechnology, Edited by Jain, P. K.,V. K. Gupta and V Bajpai. pp. 6-23. LAP Lambert Academic Publishing AG and Co. KG, Germany.

Jain, P.K., Gupta, V.K., Gaur, R.K, Bajpai, V., Gautam, N. and Modi, D.R. 2010c. Fungal enzymes: potential tools of environmental processes. In Fungal Biochemistry and Biotechnology, Edited by V. K. Gupta, Maria Tuohy and R.K.Gaur, pp 44-56. LAP Lambert Academic Publishing AG and Co. KG, Germany.

Jain, P.K., Gupta, V.K., Pathak, H., Lowry, M. and Jaroli, D.P. 2010a. Characterization Of 2t Engine Oil Degrading Indigenous Bacteria, Collected From High Altitude Mussoorie Region Of India. *World. Journal of Microbiology and Biotechnology,* 26,1419–1426.

Jain, P.K., Gupta, VK., Gaur, R.K., Lowry, M., Jaroli, D.P. and U.K. Chauhan. 2011a. Bioremediation of petroleum oil contaminated soil and water. *Research Journal of Environmental Toxicology,* 5(1), 1-26.

Jain, P.K., Lowry, M., Gupta, V.K., Bajpai, V., Sharma, S. Joshi, N.. and Gaur, R.K. 2010b. *Enterobacter* sp. as an engine oil degrader. *Biospectra,* 5(1), 115-120.

Janikowski, T., Velicogna, D., Punt, M. and Daugulis, A. 2002. Use of a two-phase partitioning bioreactor for degrading polycyclic aromatic hydrocarbons by a Sphingomonas sp. *Appl Microbiol Biotechnol.,* 59(2–3), 368-376.

Johjima, T., Ohkuma, M. and Kudo, T. 2003. Isolation and cDNA cloning of novel hydrogen peroxide dependent phenol oxidase from the basidiomycete *Termitomyces albuminosus. Appl. Microbiol. Biotechnol.,* 64, 220-225.

Johnson, K., Anderson, S and Jacobson, C.S. 1996. Phenotypic and genotypic characterization of phenanthrene degrading fluorescent Pseudomonas biovars. *Applied Environ. Microbiol.,* 62: 3818-3825.

Kadhim, H., Graham, C., Baratt, P., Evane, C.S. and Rastall, R.A. 1999. Removal of Phenolic compounds by *Coriolus versicolor* grown on wheat bran. *Enzyme Microbial Technol.,* 24, 303-307.

Kamei, I., Sonoki, S., Haraguchi, K. and Kondo, R. 2006. Fungal bioconversion of toxic polychlorinated biphenyls by white-rot fungus , *Phlebia brevispora. Appl. Microbiol. Biotechnol., 73,* 932-940.

Kanesawa, Y., Tanahashi, N., Doi, Y. 1994. Enzymatic degradation of microbial poly (3 hydroxyalkanoates). *Polym Degrad Stabil.,* 45:179–185.

Karam, J. and Nicell, J.A. 1997. Potential application of enzymes in waste treatment. J. *Chem Technol Biotechnol.,* 69, 141-153.

Kasuya, K., Doi, Y. 1994. Enzymatic degradation of poly[(R)-3-hydroxybutyrate] by Comamonas testosteroni ATSU of soil bacterium. *Polym Degrad Stabil.,* 45:379–386.

Kasuya, K., Inoue, Y., Doi, Y. 1996. Adsorption kinetics of bacterial PHB depolymerase on the surface of polyhydroxyalkanoate films. *Int. J. Biol. Macromol.,* 19:35–40.

Kasuya, K., Inoue, Y., Yamada, K., Doi, Y. 1995. Kinetics of surface hydrolysis of poly[(R)-3-hydroxybutyrate] film by PHB depolymerase from *Alcaligenes faecalis* T1. *Polym Degrad Stabil.,* 48:167–174.

Kay, M. J., Morton, L.H.G. and Prince, E.L. 1991. Bacterial degradation of polyester polyurethane. *Int. Biodeterior.,* 27, 205–222.

Kelly, R.L., Liu, B. and Srivastava, V. 1998. Landfarming: A practical guide. In: Sikdar SK and Irvine RL (Eds), Bioreme-diation: Principles and Practice-Bioremediation Technologies, Vol. 3 pp. 223–243, Technomic Publishing CO, Inc, Lancaster, Pennsylvania, USA.

Kennicutt, M.C. 1988. The effect of biodegradation on crude oil bulk and molecular composition. *Oil and Chem. Pollut.,* 4, 89-112.

Kita, K., Ishimaru, K., Teraoka, M., Yanase, H., Kato, N. 1995. Properties of poly (3-hydroxybutyrate) depolymerase from a marine bacterium, *Alcaligenes faecalis AE122. Appl. Environ. Microbiol.,* 61:1727–1730.

Kiyohara, H., N. Takizawa and K. Nagao, 1992. Natural distribution of bacteria metabolizing many kinds of polycyclic aromatic hydrocarbons. J. Ferment. Bioeng., 74: 49-51.

Koker, T.H., Burdsall, H., Jr and Janse, B.J.H. 2000. A provisional name for a taxon of *Phanerochaete* from South Africa. *Sydowia,* 52, 10-15.

Kotterman, M., Wasseveld, R. and Field, J. 1996. Hydrogen peroxide production as a limiting factor in xenobiotic compound oxidation by nitrogen-sufficient cultures of *Bjerkandera* sp. strain BOS55 overproducing peroxidases. *Appl. Environ. Microbiol.,* 62,880-885.

Lambert, M., Kremer, S., Sterner, O. and Anke, H. 1994. Metabolism of pyrene by the basidiomycete *Crinipellis stipitaria* and identification of pyrenequinones and their hydroxylated precursors in strain JK375. *Appl. Environ. Microbiol.,* 60, 3597–3601.

Lange, B., Kremer, S., Sterner, O. and Anke, H. 1994. Pyrene metabolism in Crinipellis stipitaria: identification of trans-4,5-dihydro-4,5-dihydroxypyrene and 1-pyrenylsulfate in strain JK364. *Appl. Environ. Microbiol.,* 60, 3602–3607.

Luke, A.K. and Burton, S.G. 2001. A novel application for *Pseudomonas putida* progress from batch culture to a membrane bioreactor for the bioremediation of phenols. *Enzyme Microbial Technol.,* 29, 348-356.

Madigan, M.T., Martinko, J.M., Dunlap, P.V. and Clark, D.P. 2010. *Brock Biology of Microorganisms,* 12th edition, Benjamin Cummings.

Margaret, J., Webb, A. anderson, C., Wouters, A. and Swings, J. 1993. Microbial degradation of poly (3-hydroxybutyrate) and poly (3-hydroxybutyrate-co-3-hydroxyvalerate) in soils. *Appl. Environ. Microbiol.,* 59, 3233–3238.

Margesin, R., Zimmerbauer, A. and Schinner. F. 1999. Soil Lipase activity – a useful indicator of oil biodegradation. *Biotechnol. Tech.*, 13, 859–863.

Marijke, M.A. and van-Vlerken, F. 1998. Chances for biological techniques in sediment remediation. *Wat. Sci. Tech.*, 37(6–7), 345–353.

Matsubara, M., Lynch, J.M. and Leij, F.A.A.M. De. 2006. A simple screening procedure for selecting fungi with potential for use in the bioremediation of contaminated land. *Enzyme Microb. Technol.,* 39, 1365-1372.

Mayer, A.M. and Staples, R.C. 2002. Laccase: new functions for an old enzyme. *Phytochemistry,* 60, 551-565.

Moen, M.A. and Hammel, K.E. 1994. Lipid peroxidation by the manganese peroxidase of Phanerochaete chrysosporium is the basis for phenanthrene oxidation by the intact fungus. *Appl. Environ. Microbiol., 60,* 1956–1961.

Mohan, V.S., Shailaja, S., Krishna, M.R., Reddy, K.B. and Sarma, P.N. 2006. Bioslurry phase degradation of di-ethyl phthalate (DEP) contaminated soil in periodic discontinuous mode operation: Influence of bioaugmentation and substrate partition. *Process Biochem.,* 41(3), 644-652.

Monib, M., Abd-el-Malek, Y., Hosny, I. and Fayez, M. 1979. Effect of Azotobacter inoculation on plant growth and soil nitrogen, *Zentralbl. Bakteriol. Naturwiss.,* 134, 140.

Morgan, P. and Watkinson, R.J. 1989. Hydrocarbon degradation in soils and methods for soil biotreatment, CRC Critic. *Rev. Biotechnol.,* 8(4), 305–332.

Morton, L.H.G.. and Surman, S.B. 1994. Biofilms in biodeterioration—a review. *Int. Biodeterior. Biodegrad.,* 32, 203–221.

Mueller, J.G., Lantz, S.E., Blattman, B.O. and Chapman, P.J. 1991. Bench-scale evaluation of alternative biological treatment processes for the remediation of pentachlorophenol and creosote contaminated materials: slurry-phase bioremediation. *Environ. Sci. Technol.,* 25, 1045-1055.

Mukai, K., Yamada, K., Doi, Y. 1993. Kinetics and mechanism of heterogeneous hydrolysis of poly[(R)-3-hydroxybutyrate] film by PHA depolymerases. *Int. J. Biol. Macromol.,* 15:361–366.

Mukai, K., Yamada, K., Doi, Y. 1992. Extracellular poly (hydroxyalkanoate) depolymerases and their inhibitor from *Pseudomonas lemoignei. Int. J. Biol. Macromol.,* 14:235–239.

Mulbry, W. and Eaton, R.W. 1991. Purification and characterization of the N-methylcarbamate hydrolase from Pseudomonas strain CRL-OK. Appl. Environ. *Microbiol.*, 57, 3679-3682.

Muller, B., Jendrossek, D. 1993. Purification and properties of poly (3-hydroxyvaleric acid) depolymerase from *Pseudomonas lemoignei. Appl Microbiol. Biotechnol.*, 38:487–492.

Nair, C.I., Jayachandran, K. and Shashidhar, S. 2008. Biodegradation of phenol. African *Journal of Biotechnology*, 7(25), 4951-4958.

Nair, Indu C., Ayachandran, K. and Shankar-Shashidhar. 2008. Biodegradation of phenol, *African Journal of Biotechnology,* 7(25), 4951-4958.

Nakamura, K., Tomita, T., Abe, N. and Kamio, Y. 2001. Purification and characterization of an extra cellular poly (L-lactic acid) depolymerase from a soil isolate, Amycoatopsis sp. Strain K104-1. *Appl. Environ. Microbiol.*, 67, 345-353.

Nannipieri, P., Bollag, J. M. 1991. Use of enzymes to detoxify pesticide-contaminated soils and waters. *J. Environ. Qual.*, 20, 510-517.

Nano, G., Borroni, A. and Rota, R. 2003. Combined slurry and solid-phase bioremediation of diesel contaminated soil. *J. Hazard Mater.,* 100(1–3), 79-94.

Nicell, J.A. 2001. Environmental applications of enzymes. Interdisc. *Environ*, Rev., 3, 14-41.

Nojima, S., Mineki, S. and Iida, M. 1996. Purification and characterization of extracellular poly(3-hydroxybutyrate) depolymerases produced by *Agrobacterium* sp. K-03. *J. Ferment. Bioeng.*, 81,72–75.

Novotny, C., Erbanova P., Sasek V., Kubatova A., Cajthaml T., Lang E., Krahl J. and Zadrazil F. 1999. Extracellular oxidative enzyme production and PAH removal in soil by exploratory mycelium of white rot fungi. *Biodegradation,* 10, 159-168.

Novotny, E., Vyas, B., Erbanova, P., Kubatova, A. and Sasek, V. 1997. Removal of PCBs by various white rot fungi in liquid cultures. *Folia Microbiol.,* 42, 136-140.

Oberbremer, A., Müller-Hurtig, R. and Wagner, F. 1990. Effect of the addition of microbial surfactants on hydrocarbon degradation in a soil population in a stirred reactor. *Appl. Microbiol. Biotechnol.,* 32(4), 485-489.

Okeke, B.C., Paterson, A., Smith, J.E. and Watson, C.I.A. 1997. Comparative biotransformation of pentachlorophenol in soils by solid substrate cultures of *Lentinula edodes. Appl. Microbial. Biotechnol.*, 48, 563-569.

Oudot, J. 1984. Rates of microbial degradation of petroleum components as determined by computerized capillary gas chromatography and computerized mass spectrometry. *Mar. Env. Res.,* 13, 277-302.

Pathak, H., Jain, P.K., Jaroli, D.P. and Lowry, M. 2008. Degradation of phenantherene and anthracene by Pseudomonas strain, isolated from coastal area. *Bioremediation journal.* 12, 111-116.

Pathak, H., Vashistha, A., Jain, P.K., Nagmani, A., Jaroli, D.P., Lowry, M. 2010. Physcio-chemical properties of petroleum contaminated soil collected from coastal areas of Mumbai. *Asian Journal of Experimental Science,* 24(1), 175-178.

Pathirana, R. A. and Seal, K.J. 1984. Studies on polyurethane deteriorating fungi. Part 1. Isolation and characterization of the test fungi employed. Int. *Biodeterior.,* 20, 163–168.

Pathirana, R. A. and Seal, K.J. 1984. Studies on polyurethane deteriorating fungi. Part 2. An examination of their enzyme activities. *Int. Biodeterior.*, 20, 229–235.

Pearce, K.. and Ollermann, R.A. 1998. Status and scope of bioremediation in South Africa. In: Sikdar SK and Irvine RL (Eds) Bioremediation: Principles and Practice-

Bioremediation Technologies, Vol. 3, pp. 155–182. Technomic Publishing CO, Inc, Lancaster, Pennsylvania, USA.

Pieper, D.H. et al. 2004. Genomic and mechanistic insight into the biodegradation of organic pollutants. *Curr. Opin. Biotechnol.,* 15, 215–224.

Pinelli, D., Fava, F., Nocentini, M. and Pasquali, G. 1997. Bioremediation of a polycyclic aromatic hydrocarbon-contaminated soil by using different aerobic batch bioreactor system. *J. Soil Contam.,* 243-256.

Pointing, S.B. 2001. Feasibility of bioremediation by white-rot fungi. *Appl. Microbiol. Biotechnol.*, 57, 20-32.

Prenafeta-Boldu, X.F., Kuhn, A., Mam, L. D., Anke, H., Van Groenestijin, J.W. and De Bont, J.A.M. 2001. Isolation and characterization of fungi growing on volatile aromatic hydrocarbons as their sole carbon and energy source. *Mycol. Res.,* 4: 477-484.

Prieto, M.B, Hidalgo, A., Rodriguez, F.C, Serra, J.L. and Llama, M.J. 2002. Biodegradation of phenol in synthetic and industrial wastewater by *Rhodococcus erythropitics*. UPV-1 immobilized in air-stirred reactor with clarifier. *Appl. Microbiol. Biotechnol.,* 58, 583-859.

Prince, R.C. 1993. Petroleum spill bioremediation in marine environments. *Crit. Rev. Microbiol.*, 19, 217-242.

Ramos, H.J., Roncato-Maccari, L.D., Souza, E.M., Soares-Ramos, J.R., Hung, M. and Pedrosa, F.O. 2002. Monitoring *Azospirillum*–wheat interactions using the *gfp* and *gusA* genes constitutively expressed from a new broad-host range vector, *J. Biotechnol.,* 97, 243.

Ramos, J.L. 2005. Bioremediation of polynitrated aromatic compounds: plants and microbes put up a fight. *Curr. Opin. Biotechnol.,* 16, 275–281

Raybuck, S.A. 1992. Microbes and microbial enzymes for cyanide degradation. *Biodegradation,* 3, 3-18.

Readman, J.W., Fowler, S.W., Villeneuve, J.-P., Cattini, C., Oregioni, B. and Mee, L.D. 1992. Oil and combustion-product contamination of the Gulf marine environment following the war. *Nature*, 358, 662-665.

Reddy, C.A. 1995. The potential for white-rot fungi in the treatment of pollutants. *Curr. Opin. Biotechnol.,* 6, 320-328.

Rittmann, B.E. and Schloendorn, J. 2007. Engineering away lysosomal junk: medical bioremediation. *Rejuvenation Research*, 10(3), 359–365.

Robles, A., Lucas, R., Cienfueges, A.D. and Galvez, A. 2000. Phenol oxidase activity in strains of the hyphomycete *Chalara paradoxa* isolated from Olivemill wastewater disposal ponds. *Enzyme Microbial Technol.*, 26, 484-490.

Robles-González, I., Ríos-Leal, E., Sastre-Conde, I., Ferrera-Cerrato, R., Caffarel-Méndez, S. and Poggi-Varaldo, H.M. 2003. Effect of supplementary organic carbon source and electron acceptors on the removal of 2,4-Dichlorophenoxyacetic acid in slurry bioreactors. In Proceedings 2nd International Conference on Remediation of Contaminated Sediments, Venice, September 30, 2003; Venice, Italy Edited by: Pellei M, Porta A, Hinchee E. Battelle Press, Columbus, OH; 2003. Paper G-08.

Rodríguez, Couto, S. and Toca-Herrera, J.L. 2006. Industrial and biotechnological applications of laccases*: A review. Biotechnol. Adv.,* 24, 500- 513.

Rubilar, O., Diez, M. C. and Gianfreda, L. 2008. Transformation of chlorinated phenolic compounds by white rot fungi. Crit. *Rev. Environ. Sci. Technol.*, 38, 227-268.

Rusansky, S., Avigad, R., Michaeli, S. and Gutnick, D.L. 1987. Involvement of a plasmid in growth on and dispersion of crude oil by*Acinetobacter calcoaceticus* RA 57. *Applied Environ. Microbiol.* 53: 1918-1923.

Sack, U. and Gunther, T. 1993. Metabolism of PAH by fungi and correlation with extracellular enzymatic activities. *J. Basic Microbiol.*, 33, 269–277.

Samanta, S.K. 2002. Polycyclic aromatic hydrocarbons: environmental pollution and bioremediation. *Trends Biotechnol.*, 20, 243–248

Santos, H.F., Carmo, F.L., Paes, J.E., Rosado, A.S. and Peixoto, R.S.2011. Bioremediation of mangroves impacted by petroleum, *Water, Air and Soil Pollution*, 216(1–4), 329–350.

Sasaki, T., Maki, H., Ishihara, M. and Harayama, S. 1998. Vanadium as an internal marker to evaluate microbial degradation of crude oil. *Environ. Sci. Technol.*, 22, 3618-3621.

Schirmer, A., D. Jendrossek. and H. G. Schlegel. 1993. Degradation of poly (3-hydroxyoctanoic acid) (P(3HO)) by bacteria: purification and properties of a P(3HO) depolymerase from *Pseudomonas fluorescens* GK13. *Appl. Environ. Microbiol.*, 59, 1220–1227.

Schneider, P., Michael, B.C., Kristine, M., Torben,, Lars, K.S., Peter, R.O., Kiberly, M.B., Stephen, H.B. and Feng, X. 1999. Characterization of a *Coprinus cinereus* laccase. *Enzyme microbial Technol.*, 25, 502-508.

Service Centrum Grond (SCG). 2004. The mismatch between (in situ) soil site investigation and (ex situ) excavated soil quality. http://www.scg.nl/SCG/files/in-situ ex-situ.pdf (as on 1 July 2004).

Shashirekha, S., Uma, L. and Subramanian, G. 1997. Phenol Degradation by themarine cyanobacterium Phormidium valderianum BDU 30501. J. Ind. *Microbiol. Biotechnol.*, 19, 130-133.

Shawabkeh, R., Khaled, M. and Khleifat. 2007. Rate of biodegradation of phenol by Klebsiella oxytoca in minimal medium and nutrient broth conditions. *Bioremediation J.*, 11, 13-19.

Singh, C.J. 2002. Optimization of an extra cellular protease of Chrysosporium keratinophilum and its potential in bioremediation of keratinic wastes. *Mycopathologia*, 156, 151- 156.

Singh, Harbhajan. 2006. Mycoremediation: fungal bioremediation. New York: Wiley-Interscience.

Singh, R., Sharma, R., Tewari, N., Geetanjali and Rawat D.S. 2006. Nitrilase and its application as a 'green' catalyst. *Chem. Biodiv.*, 3, 1279-1287.

Smith, M., Thurston, F. and Wood, D.A. 1997. Fungal laccases: role in delignification and possible industrial applications, in Multi-Copper Oxi- Dases, A.Messerschmidt, ed., pp. 201–224, World Scientific Publishing, Singapore.

Steffens, L.J.C. 2002. Over expression of polyphenol oxidase in transgenic tomato plants results in enhanced bacterial disease resistance. *Planta,* 215, 239-247.

Sugano, W., Ishigaki, T., Nakanishi, A., Ike, M. and Fujita, M. 1999. Degradation of biodegradable plastics in model waste landfill rectors. *Proc. Jpn. Waste Manage. Experts,* 1073–1075 (in Japanese).

Sugiura, K., Ishihara, M. and Harayama, S. 1997. Comparison of the biodegradability of four crude oil samples by a single bacterial species and by a microbial population. *Environ. Sci. Technol.*, 31, 45-51.

Sutherland, T. D., Horne, I. and Weir, K. M. 2004. Enzymatic bioremediation: from enzyme discovery to applications, *Clinical and Experimental Pharmacology and Physiology*, 31 (11), 817–821.

Sutherland, T., Russel, R. and Selleck, M. 2002. Using enzymes to clean pesticide residues. *Pestic. Outlook,* 13, 149-151.

Tabak, H.H. and Govind, R. 1997. Bioavailability and biodegradation kinetics protocol for organic pollutant compounds to achieve environmentally acceptable end-points during bioremediation. *Ann. NY Acad. Sci.*, 829(1), 36-61.

Tanio, T., Fukui, T., Shirakura, Y., Saito, T., Tomita, K., Kaiho, T. and Masamune, S. 1982. An extracellular poly(3-hydroxybutyrate) depolymerase from *Alcaligenes faecalis. Eur. J. Biochem.,* 124, 71–77.

Thurston, C.F. 1994. The structure and function of fungal laccases, *Microbiology*, 140: 12-26.

Tien, M. 1987. Properties of ligninase from Phanerochaete chrysosporium and their possible applications. *CRC Crit. Rev. Microbiol.*, 15, 141-168.

Tilawat, A.K.. and Jain, P.K. 2011. Microbial bioremediation. In Recent Advances in Environmental Biotechnology, Edited by Jain, P.K., V.K. Gupta and V. Bajpai. pp. 170-188. LAP Lambert Academic Publishing AG and Co. KG, Germany.

Torres, L.G, Alicia-Sanchez, de-la-vega, Norma, A. and Jimmenz, K.B. 1998. Production and characterization of calcium alginate biocatalyst for removal of phenol and chlorophenols from waste waters. *Proc. Biochem.*, 33(6), 625-634.

van-Hamme, J.D., Singh, A. and Ward, O.P. 2003. Recent advances in petroleum microbiology, *Microbiology and Molecular Biology Reviews*, 67(4), 503–549.

Vyas, B. R. M., Bakowski, S., Sasek, V. and Matucha, M. 1994.Degradation of anthracene by selected white rot fungi. *FEMS Microbiology Ecology*, 14(1):65–70.

Wang, S.Y. and Vipulanandar, C. 2001. Biodegradation of naphthalene-contaminated soils in slurry bioreactors. *J. Environ. Eng.*, 127(8),748-754.

Wang, Z., Fingas, M., Blenkinsopp, S., Sergy, G., Landriault, M., Sigouin, L., Foght, J., Semple, K. and Westlake, D.W.S. 1998. Comparison of oil composition changes due to biodegradation and physical weathering in different oils. *J. Chromatography,* 809, 89-107.

Wenk, M., Baumgartner, T., Dobovsek, J., Fuchs, T., Kucsera, J., Zopfi, J. and Stucki, G.1998. Rapid atrazine mineralisation in soil slurry and moist soil by inoculation of an atrazine-degrading *Pseudomonas* sp. strain. *Appl. Microbio.l Biotechnol.,* 49(5), 624-630.

Williams, R.T. and Keehan, K.R. 1993. Hazardous and industrial waste composting. In Science and Engineering of Composting. Hoitink, H.A.J. and Keener, H.M., Eds. pp 363–382. Renaissance Press, Worthington.

Woo, S.H. and Park, J.M. 1999. Evaluation of drum bioreactor performance used for decontamination of soil polluted with polycyclic aromatic hydrocarbons. *J Chem Technol Biotechnol.,* 74(10), 937-944.

Wu, Y., Keith, F.T., Nihar, B. and Jatinder, K.B. 1998. A model for the protective effect of additives on the activity of horseradish peroxidase in the removal of phenol. *Enzyme Microbial Technol.,* 22, 315-322.

Xiangchun, Q. and Zhang, Y.M. 2003. Biodegradation of 2,4 dichloro-phenol in an airlift honeycomb like ceramic reactor. *Proc. Biochem.*, 38, 1545-1551.

Youn, H.D., Kim, K.J., Maeng, J.S., Han, Y.H., Jeong, I.B., Jeong Kang, G. S.O. and Hah, Y.C. 1995. Single electron transfer by an extracellular laccase from the white-rot fungus *Pleurotus ostreatus*. *Microbiology,* 141, 393–398.

Yue, C.L., Gross, R.A. and McCarthy, S.P. 1996. Composting studies of poly (b-hydroxybutyrate-co-b-hydroxyvalerate). *Polym. Degrad. Stab*., 51, 205–210.

Yukie-Akutsu, Toshiaki-Nakajima-Kambe,Nobuhiko Nomura and Tadaatsu Nakahara. 1997. Purification and Properties of a Polyester Polyurethane-Degrading Enzyme from *Comamonas acidovorans* TB-35; *Applied and Environmental Microbiology,* 64 (1), 62–67.

Zach, A., Latif, M., Binner, E. and Lechner, P. 1999. Influence of mechanical-biological pretreatment on the toxicity of municipal solid waste. *Compost Sci. Util.,* 7(4), 25–3.

Zahida-Deva, W., Gulam, M., Peerzuda, M.D. and Diagambar, V.B. 1998. Oxidation of phenols by horseradishperoxidase and lactoperoxidase compound II. *Kinetic consideration,* 35, 353-357.

Zhang, C., Hughes, J.B, Nishino, S.F. and Spain, J.C. 2000. Slurry-phase biological treatment of 2,4-dinitrotoluene and 2,6-dinitrotoluene: role of bioaugmentation and effects of high dinitrotoluene concentrations. *Environ. Sci. Technol*., 34(13), 2810-2816.

In: Molecular Biology of Bacteria
Editors: R. Kumar Gaur and H. K. Gautam
ISBN: 978-1-62618-189-2
© 2013 Nova Science Publishers, Inc.

Chapter 7

Nanocatalysis: A Prospective and Sustainable Approach for the Chemical World

Kashyap Kumar Dubey *** *and Dhirendra Kumar***
Microbial Biotechnology Laboratory,
University Institute of Engineering and Technology, M.D. University Rohtak,
Rohtak, India

Abstract

This chapter illustrates a prospective overview of the nanoparticles and use of nanocatalyst, in the health care sector, particularly. The market for nanotechnology based products is growing day-by-day and by the end of 2015 about $ 29.7 billion will reached. Nanocatalysis involves in ample areas, from fundamental research to commercial applications in food, pharmaceutical, and energy. It has emerged as sustainable alternatives to conventional processes, due to robust high surface area for catalysis. The nanosized particles increase the uncovered surface area of the active component of the catalyst, by this way improving the contact between substrate and catalyst, considerably. The nanocatalysts are inexpensive, stable, easily recycled and reused for several batches with reliable activity.

Introduction

Nature is exceptionally diverse in terms of the large number of organic molecules required by living world. This diversity is possible due to the wide-range of catalysis by the enzymes. Since, ancient period enzymes have been used, without any scientific knowledge, to preserve food, and to prepare curd. Many of these old processes have survived, and been commercialized such as, beer making and the production of milk products [2,3,4]. Substantial

* Corresponding author Email: kashyapdubey@yahoo.com.

research has focused in 1960's on large-scale application of immobilized enzymes to optimize the structure of matrix/carrier materials for better catalytic efficiency.

Table 1. Commercially manufactured Nanoparticles

Type	Applications
Carbon nanotubes (Single-wall and Multiwall nanotubes)	Additives for polymer composites, Electronic field emitters, Batteries, Fuel Cell
Fullerenes (C_{60}) [8]	Mechanical and tribological applications
Metal oxides (TiO_2 , SiO_2, Al_2O_3, ZrO_2, ZnO_2)	Additives for polymer composites, UV-A protection, Solar cells, Pharmaceutical, additives for scratch resistance coatings
Compound semiconductors (GaAs, CdTe)	Electronic and optical devices
Organic nanoparticles	Micronised drugs and chemicals (vitamins, pigments, pharmaceuticals), polymer dispersion
Metals (Ag, Au and Ni)	Catalytic applications, Optoelectronics, would dressing

In the today's world, nanotechnology has fetched a revolution in numerous areas such as health, food, energy, environmental protection, and many more areas. Among these areas, connecting catalytic processes by using nano-science remains one of the key areas of research because of its direct impact on human being and its development. Besides, sustainable catalytic approaches, it can enhance our value of living, improve healthiness and amplify our life expectancy, and additional sustainable energy [1,9,11] . Catalysis enhances numerous chemical processes, from academic research laboratories to the chemical industry. Variety of products, such as medicines, fine chemicals, fuels, paints, lubricants, and other value added products essential to humans, would not be feasible in the absence of catalysts [15,19]. There are various types of nanoparticles which are commercially manufactured given in Table 1.

The recent developments in the field of biocatalyst have enabled the biological processes to compete successfully with old chemical processes. In many cases, a bio-based process is the most straightforward and economical way to produce these products [11,19,21,22]. However, in all biological processes, the synthesis of the desired product is catalyzed by enzymes. There are examples where biocatalysts may compete with traditional chemical processes. In view of above the raw materials' availability and cost, nanocatalytic route have replaced the chemical routes. Recently, researchers have taken keen interest on the development of catalytic systems due to their increasing importance in organic chemistry [11,19]. Previously, heterogeneous catalysis is being used in the fine chemicals industries but now they need for more eco-friendly production process due to the associated problems (Figure 1). This trend is assisted by the availability of novel catalytic materials and modern techniques of generating and exploring specific active sites on catalyst surfaces.

In homogeneous catalyst, the catalyst is in the same phase as the substrate, is generally accepted by chemists. They are widely used in several of commercial applications, but due to the difficulty of catalyst separation from the final product creates economic and environmental issues to broadening their scope (Figure 1). Removal of trace amounts of catalyst from the end product is essential since metal contamination is highly regulated, in the pharmaceutical industry by Food and Drug Administration (FDA).

Therefore, new catalyst system requires that allow for rapid, selective chemical transformations with high product yield [18,20] and easy catalyst separation and recovery are highly desired for eco-friendly manufacturing processes.

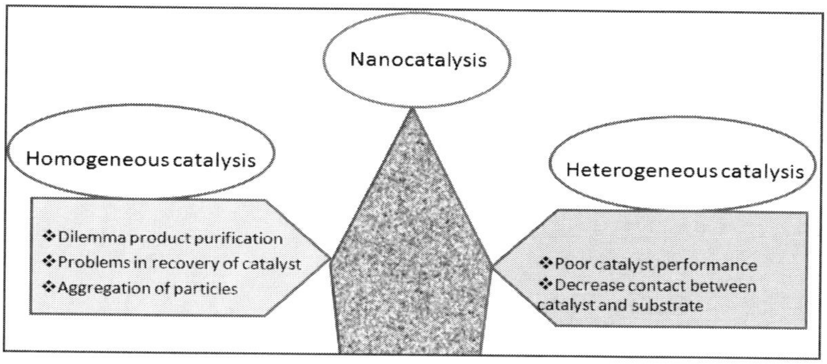

Figure 1. Nanocatalysis a bridge between homogenous and heterogeneous catalysis.

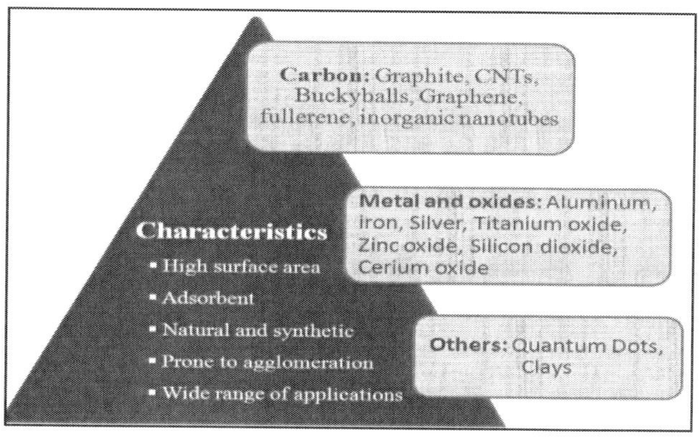

Figure 2. Various types of Nanomaterials used for nanocatalysis.

What is Nanocatalysis?

In Nanocatalysis, enzymes are incorporated into nano-structured materials like nano-porous media, nano-fibers, carbon nano-tubes and other nanoparticles, which have evident immense efficiency in the exploitation of the nano-scale surroundings of the enzyme.

Besides this, nano-scale materials provide the better confines in correspond to the important factors that determine the efficiency of biocatalysts; which includes surface area, effective enzyme loading and mass transfer (Figure 2).

Applications of Nanocatalysis

The size of catalyst particles has a significant effect on catalysis including, their shape and morphology. In the recent past, biocatalyst (free enzymes) and immobilized biocatalyst was used for stereospecific reactions in chemical industry, due to certain limitations and poor yield [7, 18]; scientists have already been demonstrated use of nanocatalyst with specific reaction and high yield (Figure 3) [16,17]. Besides, reactivity and stability of the nanocatalyst, selectivity was also influenced to commercial manufactures.

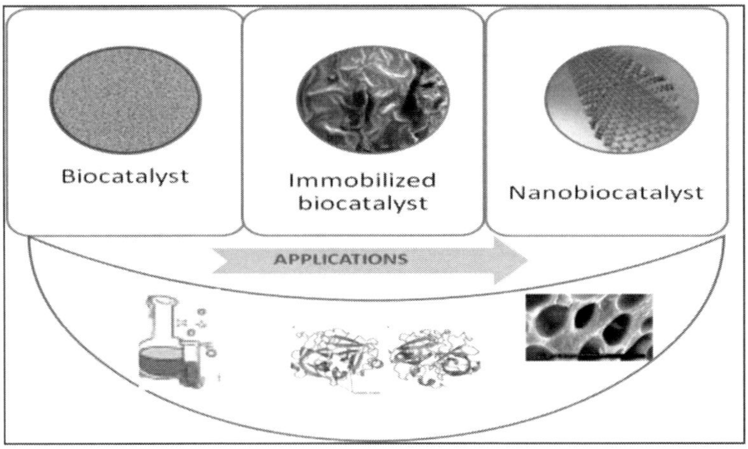

Figure 3. Applications of various catalysts in sequential order.

Problems Associated with Biocatalysts

Though, the high promises of the use of biocatalysts in chemical industry still limited [5, 6, 14] due to the following reasons:

- a limited knowledge of biocatalyst mechanisms
- poor understanding of metabolic pathways for secondary metabolites, including pathway interactions
- a limited number of methods to engineer whole organisms, i.e., metabolic engineering
- the high cost of producing many enzymes and co-factors for biocatalyst application

Global RandD Trends

Presently, it is difficult to expect the specific long term global RandD trends of the nanotech industry. Various types of nanoparticles are being produced in large quantities by many industries (Table 2). Realizing, that the future trends are depends on the relation of many complex factors which are difficult to predict, particularly for growing industry. [10,12,13]. Research and Development trends can evidently be expected to emerge in the following areas specifically:

- Health sector: cancer therapeutics, drug delivery, imaging
- Energy sector: hydrogen storage, efficiency improvement, catalysis

- Agriculture sector: increased crop yields, secure packaging
- Environment sector: water filtration, remediation, reduced air emissions, chemical and biological sensing.

Table 2. Present status of Nanoparticles manufacturers' world-wide

Name of the manufacturer	Material produced
Abraxis Biosciences, Inc. USA	Chemotherapeutic medications as ABI007, which encapsulate paclitaxal to reduce toxicity
Altair Nanotechnologies, Inc. USA	Nanomaterials for energy and life sciences applications
Artimplant AB, Sweden	Devices for orthopedic surgery (biodegradable materials)
Auto Fibre Craft, India	Nano-size Silver Powder for use in electronic applications (making conductive inks and pastes)
Bayer AG ,Germany	Therapeutics as well diagnostics devices specifically for cancer , and variety of applications
Bee Chems, India	Nano Silica products
Bilcare, India	nonClonable: unique security technology
Cell Robotics International Inc. USA	Expert in biophonotic technologies for medical and clinical research
Dabur Pharma, India	Novel drug delivery system for Paclitaxel
Kerala Minerals and Metals, India	Titanium dioxide nanoparticles
MIV Therapeutics Inc. Canada	Expert in stent coatings and therapeutics for stent based drug delivery systems
Mitsubishi Corporation, Japan	Large-scale fullerene
Nanoshel, India	Nanotubes, SWCNT´s, MWCNT´s, nanoparticles
Nucryst Pharmaceuticals,Canada	Expert in wound care and pharmaceuticals products based on silver nanocrytalline technology
Proteome Systems,Australia	Manufacturer of Rapid Diagnostic Kits for Tuberculosis
Quantum Materials cop., India	Carbon nanotubes and grapheme
Schering AG, Germany	Expert in therapeutics for Oncology, and diagnostics tool for MRI
Silvidia Limited, Australia	Biomedical Technologies for specialized silicon products ,for human and animal healthcare
Sisco Research Laboratories (SRL), India	Nanopowders and Carbon Nanotubes
Starpharma Group, Australia	Expert in a variety of polyvalent compound, focused on developing anti infective therapeutics
Straumann AG, Switzerland	Expert in Dentistry and Dental regeneration products
United Nanotechnologies, India	Nanoparticles-based coatings

Conclusion

Catalysts are a crucial part of the modern chemical industry, and are used in a huge number of chemical processes all over the world. Researchers are constantly pushing ahead to improve the performance and lifetime of nanocatalysts, by which the profitability of much chemical manufacturing revolve on. Nanomaterials in catalysis are offering a whole new array of parameters which scientists can squeeze to try and find the perfect catalyst support. As this research goes on, we understand more and more about how catalysts behave at the nanoscale, and in future we will able to make chemicals ever more efficiently.

References

[1] D., Dwivedi, and K. Gopal. (2010). Biosynthesis of silver and gold nanoparticles using Chenopodium album leaf extract colloids and surface *A, Physicochem Eng. Aspects*, 36,927-33.

[2] Aitken RJ, Creely KS, Tran CL. (2006). Manufacture and use of nanomaterials: current status in the UK and global trends , *Occup. Med.*, 56 ,5, 300-306.

[3] Alivisatos, A. P. (1996). Semiconductor clusters, nanocrystals, and quantum dots. *Science*, 271, 933 937.

[4] Benkovic, S.J. and A. Ballesteros (1997). Biocatalysts the next generation, *TIBTECH*, 15, 385-386.

[5] Chaudhry Q, Boxall A, Aitken R, Hull M. A (2005). Scoping Study into the Manufacture and Use of Nanomaterials in the UK. Sand Hutton, York: Central Science Laboratory.

[6] Fabian Käsermann, and Christoph Kempf (1998), Buckminsterfullerene and Photodynamic Inactivation of Viruses , *Rev. Med. Virol.*, 8, 143–151

[7] Huczko A, Byszewski P. Fullerines. (1998). Characteristics of Production Hygiene Review. HSE Research Report 274. London: HSE Books.

[8] Iijima S. (1991) Helical microtubules of graphitic carbon. *Nature*, 354, 56–58.

[9] Jortner J, Rao CNR. (2002). Nanostructured advanced materials. Perspectives and directions. *Pure Appl. Chem.*, 74,1491–1506.

[10] Kerker, M. (1985). The optics of colloidal silver: something old and something new. *J. Colloid Interface Sci.*, 105, 297–314.

[11] Liu, Z. Robinson, J. T.,Sun, X. M.,Dai, H. J. (2008) PEGylated nanographene oxide for delivery of water-insoluble cancer drugs. *J. Am. Chem. Soc.*, 130, 10876-10877.

[12] Marshall, C.T. and J.M. Woodley (1996). Process Synthesis for Multiple-Step Microbial Conversions, *Biotechnology*. 13, 1072-1078.

[13] May, S.W. (1997). New applications of biocatalysis, *Current Opinion in Biotechnology*, 8,181 - 186.

[14] Maynard AD, Baron PA, Foley M, Shvedova AA, Kisin ER, Castranova V. (2004). Exposure to carbon nanotube material: aerosol release during the handling of unrefined single wall carbon nanotube material. *J. Toxicol. Environ. Health*, 67,87–107.

[15] Murray, C. B.,Norris, D. J., Bawendi, M. G.,(1993). Synthesis and characterization of nearly monodisperse CdE (E = S, Se, Te) semiconductor nanocrystallites. *J. Am. Chem. Soc.*, 115, 8706 -8715.

[16] Park, J., Joo, J., Kwon, S. G.,Jang, Y.Hyeon, T.(2007). Synthesis of monodisperse spherical nanocrystals. *Angew. Chem. Int. Ed.,* 46, 4630-4660.

[17] Rao CNR. (2004). New developments of nanomaterials. *J. Mat. Chem.*, 4, 437–780.

[18] Rao CNR, Muller A, Cheetham AK, (2004).The Chemistry of Nanomaterials. Synthesis, Properties and Applications. Weinheim, Germany: Wiley-VCH.

[19] S. P. Dubey , M. Lahtinen, M. Sillanpaa (2010).Tansy fruit mediated greener synthesis of silver and gold nanoparticles. *Process Biochemistry*, 45, 7,1065–1071.

[20] Shi, J.J. , Zhu, Y.F , Zhang, X.R , Baeyens, W.R.G., Garcia-Campana AM.(2004). Recent developments in nanomaterials optical sensors. *Trends Anal Chem.*, 233,351–360.

[21] Xia, Y., Yang, P., Sun, Y., Wu, Y., Mayers, B., Gates, B.,Yin, Y.; Kim, F.; Yan, H.(2003). One-dimensional nanostructures: Synthesis, characterization, and applications. *Adv. Mater.* 15, 353- 389.

[22] Yin, M., O'Brien, S. (2003). Synthesis of monodisperse nanocrystals of manganese oxides. *J. Am. Chem. Soc.*, 125, 10180- 10181.

In: Molecular Biology of Bacteria
Editors: R. Kumar Gaur and H. K. Gautam
ISBN: 978-1-62618-189-2
© 2013 Nova Science Publishers, Inc.

Chapter 8

Probiotics and Herbs: Emerging Treatment Options for Non Alcoholic Fatty Liver Diseases

Chand Ram* and Vijay Kumar
Bioremediation Laboratory, Dairy Microbiology Division,
National Dairy Research Institute, Karnal-132001, Haryana, India

Abstract

Non alcoholic fatty liver disease (NAFLD) is characterized by fat accumulation in hepatocytes at or above 5% of liver weight in the absence of excessive alcohol consumption (<20 g/day). It is often associated with metabolic disorders like type 2 diabetes, obesity, and cardiovascular disease. Recently, it has become the most common chronic liver disorders with potential for causing liver failure. During last two decades, its prevalence has increased like epidemics of obesity and diabetes. NAFLD is a spectrum of disease that includes hepatic steatosis, nonalcoholic steatohepatitis (NASH), fibrosis and cirrhosis. Emerging data suggest that NASH may progress to cirrhosis and hepatocellular carcinoma (HCC) in 10-15% of patients thereby increase liver-disease related morbidity and mortality.

Although it is clinically asymptomatic, however, alterations in liver enzymes level are considered as indicator for the onset of disease. Liver biopsy is considered to be gold standard for diagnosis of the disease. Other techniques like ultrasonography, liver enzyme estimation do exist that are not much reliable. Present treatment options target the metabolic disorders such as gradual weight loss and/or use of insulin sensitizing and lipid lowering drugs. Since treatment options are limited, prevention is the right strategy in the management of NAFLD. Recent studies suggest alternate options like use of herbal products and probiotics for improvement in disease condition. Milk system is more efficient for delivery of probiotics as it favors growth and multiplication as well as protects them against harsh gastrointestinal environment.

* Corresponding author's Email address: chandram_r@yahoo.com; dmcaft2011@gmail.com.

Currently there is no recommendation for administration of probiotics for treatment or prevention of hepatological disorders. According to the World Gastroenterology Organisation Practice guidelines, probiotics can be administered only for hepatic encephalopathy.

However, administration of probiotics for treatment of non-alcoholic steatohepatitis and in prophylaxis of infections, or some complications in patients with liver cirrhosis, seems to be expected future potentials. Only a few preclinical investigations have been targeted to evaluate the efficacy of probiotics in animal models. There is hardly any proven probiotic dairy food available in the market for the treatment/ management of NAFLD.

The effect of probiotic dahi on expression of genes involved in fatty acid metabolism during NAFLD has not been investigated yet. Therefore, the present chapter will focus on the use of probiotics and their different aspects for the prevention/treatment of NAFLD.

1. Introduction

Rapidly changing life style has significantly altered food behavior of individuals by increasing the dependency on high cholesterol rich fast foods. This causes suffering of 30-40% population with various metabolic disorders like obesity, type two diabetes, and metabolic syndromes over the globe. Nonalcoholic fatty liver disease (NAFLD) is the most common cause of chronic liver disease with potential of liver failure or even hepato-cellular carcinoma in human population.

The mildest form of NAFLD is simple steatosis which is self limiting, if progresses, converts to non alcoholic steatosis (NASH) and more severe form called cirrhosis.

The term 'nonalcoholic steatohepatitis' (NASH) was introduced by Ludwig et al.,(1980), to describe an 'alcoholic-like liver disease that develops in people who do not consume significant amount of alcohol. NAFLD is the condition of fat accumulation in hepatocytes in absence of excessive alcohol consumption (<20g/day) and any other cause of hepatic steatosis. It consists a cluster of disorders *viz.*, insulin resistance, obesity, altered glucose regulation, high triglyceride and low HDL cholesterol levels, high blood pressure, hyperlipidemia, cirrhosis etc (Nobili et al., 2006).

2. Prevalence

2.1. Global

The disease has paralleled epidemics of obesity, diabetes and other metabolic syndromes during last two decades. Emerging data suggests that non-alcoholic steatohepatitis (NASH) may progress to cirrhosis, thus increasing liver-disease related morbidity and mortality. It has reached to epidemic proportion (14-15%) even among populations typically considered at low risk, especially Japan and China (Bellentani et al., 2009). The prevalence of NAFLD throughout the globe is summarized in Table 1.

Studies using aminotransferases (Alanin aminotransferases and Aspartate aminotransferase) as biomarker showed that 3-6% population suffers with NAFLD (McCullough et al., 2004). Hepatic ultrasound, a more sensitive procedure than liver function tests for detecting liver fat. It has consistently showed 20-30% prevalence of NAFLD in Western countries (Itoh, 1987; Bellentani, et al., 2000; Bellentani, et al., 2004).

Table 1. Global prevalence of nonalcoholic fatty liver disease

Country	Experimental size (cases examined)	Diagnosis method	NAFLD (%)	References
China	3543	Ultra sound	15	Zhou et al., (2007)
India	100 (Type II diabetes patients)	Ultra sound	49	Gupta et al., (2004)
India	1230	Autopsy	15.8	Amrapurkar et al., (2007)
India	541	Biochemical	35% (M), 29% (F)	Mohan et al.,(2009)
Italy	2839(Type II diabetes patients)	Ultra sound	69	Targher et al., (2007)
Japan	3432	Ultra sound	13.4	Omagari et al., (2002)
Japan	1950	Ultra sound	29	Jimba et al., (2005)
Korea	768	Ultra sound	23.4	Kim et al., (2004)
Poland	70	Biopsy	30	Halon et al., (2006)
United States	5724	ALT	2.8	Ruhl et al., (2003)
United States	88	Ultra sound	54.5	Gambarin et al., (2007)

Investigations using magnetic resonance spectroscopy to measure hepatic triglyceride content revealed that one third of United States population has excessive liver fat. Males and certain ethnic groups (45% Hispanic) tend to be at high risk (Adams et al., 2005; Shen et al., 2003).

The difference of prevalence in different ethnic groups may be ascribed to different rate of metabolic syndrome, genetic and environmental factors. Incidences of disease increase with advancement of age i.e. children 2.6 and adults 26%, respectively (Ruhl et al., 2003; Clark et al., 2003; Chin et al., 2008).

2.2. India

The disease is prevalent in 24% of general population in India and is rising at a very fast pace (Gokulakrishnan et al., 2010). Arun et al.(2006) reported NASH almost twice in obese men as compared to women (60.3% vs. 30.9%).

In a case study of 541 subjects, the fatty liver was observed in 23, 33 and 55% with normal glucose tolerance, prediabetes (impaired glucose tolerance and impaired fasting glucose) and diabetes population, respectively from Chennai, India, (Mohan et al., 2009).

3. Risk Factors

The risk factors for high incidence of NAFLD include use of drugs, parenteral nutrition, gastric bypass surgery and inherited metabolic disorders such as obesity, dyslipidemias, type 2 diabetes and insulin resistance (Tarantino et al., 2007). Several other factors have also been identified as predictors for the development of progressive fibrosis and cirrhosis. These comprise: age, hypertension, aspartate aminotransferase (AST)/alanine aminotransferase (ALT) ratio, triglycerides, iron, extent of steatosis, and inflammation (Walnes et al., 1990; Contos et al., 2001). Among non obese and non diabetic populations, the risk factors for NAFLD are; impaired fasting glycemia, hypertriglyceridemia, hyperuricemia, hypertension and low levels of HDL cholesterol (Kim et al., 2004).

3.1. Insulin Resistance and NAFLD

Insulin resistance is a major feature characterized by reduction in insulin sensitivity in hepatic and adipose tissues. The mechanisms underlying the accumulation of fat in the liver include excess dietary fat intake and increased delivery of free fatty acids to the liver, inadequate fatty acid oxidation as well as increased *de novo* lipogenesis. Insulin resistance occurs in liver primarily due to reduction in insulin-stimulated glucose transport and glycogen synthesis. Lipid accumulation within liver inhibits tyrosine phosphorylation of insulin receptor substrate-2 (IRS-2) which in turns inhibits binding and activation of phosphatidylinositol 3-kinases (PI 3-kinase) (Savage et al., 2005). Thus, the complete signaling pathway responsible for glucose uptake and triglyceride removal from liver is disturbed and as a result hyperglycemia and hyper-insulinemia get induced. Consequently, hyperinsulinemia determines an increased hepatic synthesis of free fatty acids (FFAs) and decreased synthesis of apolipoprotein B-100. This leads to triglyceride accumulation in liver and enhanced fat deposition in hepatocytes, responsible for disease condition (Charlton et al., 2002; Herruzo et al., 2006).

4. Pathogenesis

The disease is closely linked to obesity, insulin resistance, and metabolic syndrome, may also occur in non-obese patients (Clouston et al., 2004; Bacon et al., 1994).

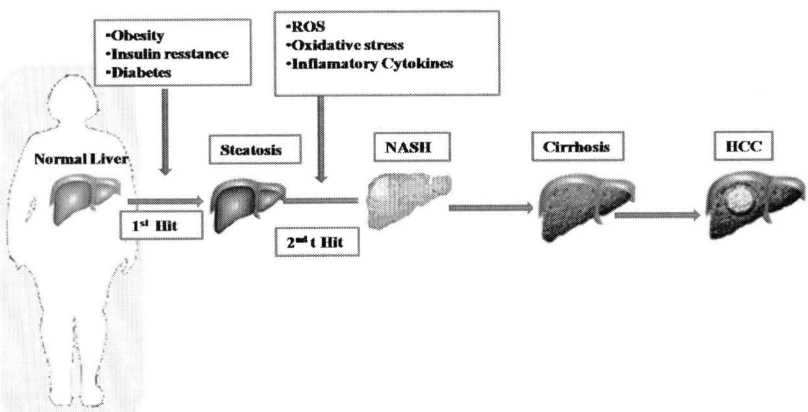

Figure 1. Multiple hit hypothesis explaining the hepatic steatosis and its progression from NAFLD to NASH, cirrhosis and hepatocellular carcinoma.

In most cases, insulin resistance develop as a result of increased insulin levels in circulation (Pinto et al.,1996). Various mechanisms of pathogenesis have been proposed, among these "multiple hit" hypothesis (Figure 1) is highly acceptable that involve insulin resistance, oxidative stress, apoptosis and adipokines like adeponectin, leptin, TNF α and TGF β (Kim et al., 2008). The first hit involves development of hepatic steatosis due to insulin resistance. The second "hit" involves role of various cytokines like plasma adiponectin, a "protective adipokine," produced from adipocytes play significant role in regulating lipid influx to liver (Pagano et al., 2005). Insulin resistance and visceral obesity lead to low level of adiponectin production that inhibits liver gluconeogenesis and suppresses lipogenesis. This results in imparement of fatty acid oxidation and increased fat accumulation in the liver.

4.1. Symptoms

Non alcoholic fatty liver disease is usually asymptomatic, although fatigue and discomfort in the right upper quadrant of the abdomen may present. The majority of patients (56–79%) are overweight with body mass index (BMI) of >25 kg/m^2. Apart from hepatomegaly, physical signs are generally absent unless cirrhosis occurred (Marchesini et al., 2003). Liver enzyme levels in some NAFLD patients fluctuate, however it remains normal in about 78% of patients. The levels are elevated, from mild to moderate and often restricted to one or both i.e. alanine aminotransferases and aspartate aminotransferase (Ipekci et al., 2003). The AST:ALT ratio is usually <1, however it may reverse in presence of cirrhosis (Sorbi et al., 1999). Alanin aminotransferase characteristically exceeds aspartate aminotransferase level i.e. ALT/AST >1 (Angulo et al., 1999). Although steatosis can be detected by non-invasive imaging, steatohepatitis can reliably be diagnosed by liver biopsy.

4.2. Causes of Triglyceride Accumulation

Hepatic steatosis is manifestation of excessive triglyceride accumulation in the liver due to excessive import of free fatty acids from adipose tissue and its minified export (secondary to reduced synthesis or secretion of VLDL) from hepatocytes, or impaired β-oxidation. The major sources of triglycerides are fatty acids stored in adipose tissue and newly synthesized in the liver through *de novo* lipogenesis (Donnelly et al., 2005).

- Impairment of VLDL synthesis and its secretion due to lipoproteinemia, protein malnutrition, or choline deficiency (Fujita et al., 2009). Patients with NASH may have a defect in postprandial ApoB secretion, leading to triglyceride accumulation (Musso et al., 2003). In addition, defect in the lipidation of Apo B, as a result of inhibition of microsomal triglyceride transfer protein (MTP) is the key mechanism in drug-induced NAFLD, e.g. amiodarone and tetracycline (Lettéron et al., 2003).
- Activation of peroxisome proliferator-activated receptor alpha (PPARα) that plays central role in stimulating β-oxidation and disposing hepatic fatty acids in NASH (Ip et al., 2003). Compromised hepatic ATP homeostasis may predispose to injury from

other insults. Adiponectin, a fat derived hormone, also have a pivotal role in increased fatty acid oxidation and decreased fatty acid synthesis (Xu et al., 2003).

- Triglyceride synthesis and oxidation are regulated by enzyme acyl-CoA and diacylglycerol acyltransferase 1 (DGAT1). DGAT1 deficient hepatocytes get protection from hepatic steatosis due to reduction in synthesis and increased oxidation of fatty acids. DGAT1 activity is necessary for the manifestation of hepatic steatosis (Villanueva et al., 2009).
- Micro RNA (MiRNA) also has a role in hepatocellular lipid metabolism and immunity. These are small non-coding RNAs that post-translationally modulate gene function and involved in cellular processes, *viz.*, cellular proliferation, inflammation, apoptosis (Bala et al., 2009). Alterations in MiRNA activity result in hepatocellular injury, apoptosis, and portal fibrosis (Hand et al., 2009). In NAFLD, specific MiRNA appear to regulate genes involved in fatty acid biosynthesis. Antagonism of specific MiRNA leads to decreased fatty acid synthesis and increased hepatic fatty acid oxidation (Jin et al., 2009).

4.3. Molecular Mechanism and Putative Targets for NAFLD

The molecular mechanism of pathogenesis of hepatic fat accumulation is poorly understood. However, studies in last decade on animal models of NAFLD have provided new insights into the molecular and physiologic alterations responsible for induction and progression of the disease. Greco et al., (2008) investigated expression of following genes associated with fatty acid metabolism in human liver with extreme steatosis due to NAFLD using Affymetrix Gene Chips.

- Glycolysis/gluconeogenesis: Dihydrolipoamide dehydrogenase (DLD), Glucose phosphate isomerase (GPI), Phosphorylase kinase alpha 2 (PHKA2) in liver etc.
- Lipid metabolism: Very low density lipoprotein receptor (VLDLR), Apolipoprotein A-V (APOA5), Apolipoprotein C-IV (APOC4), Lipase hormone-sensitive (LIPE) etc.
- Insulin pathway: Acetyl-coenzyme A carboxylase β (ACC), Protein tyrosine phosphatase receptor type F (PTPRF), Suppressor of cytokine signaling 1 (SOCS1) and Phosphorylase kinase alpha 2 (PHKA2).

Unregulated enhancement of *de novo* fatty acid synthesis is a primary disorder in NAFLD. Fatty acid synthesis by Acyl CoA carboxylase (ACC) and fatty acyl synthase (FAS) is tightly regulated by sterol regulatory element binding protein (SREBP-1c) and its expression is also negatively regulated by AMP-activated protein kinase (AMPK) and positively by insulin (Kohjim et al., 2007). In addition, increase in *de novo* fatty acid synthesis and fatty acid uptake from serum also contributes its accumulation (Figure 2). Adipose differentiation-related protein (ADRP) plays a role in fatty acid transport, although its function is not fully understood (Chang et al., 2006). Oxidation of fatty acids occurs mainly in the mitochondria. Carnitine palmitoyltransferase 1a (CPT1a) is a membrane transporter (acyl-CoA) that transports fatty acids from cytoplasm to inner mitochondrial matrix. Primarily it is a regulatory enzyme involved in mitochondrial β-oxidation. Its expression is downregulated in NAFLD due to an increase in malonyl-CoA (McGarry et al., 1977).

5. Diagnosis

Liver biopsy is the gold standard for diagnosis and differentiation of NASH from steatosis with or without inflammation in NAFLD patients.

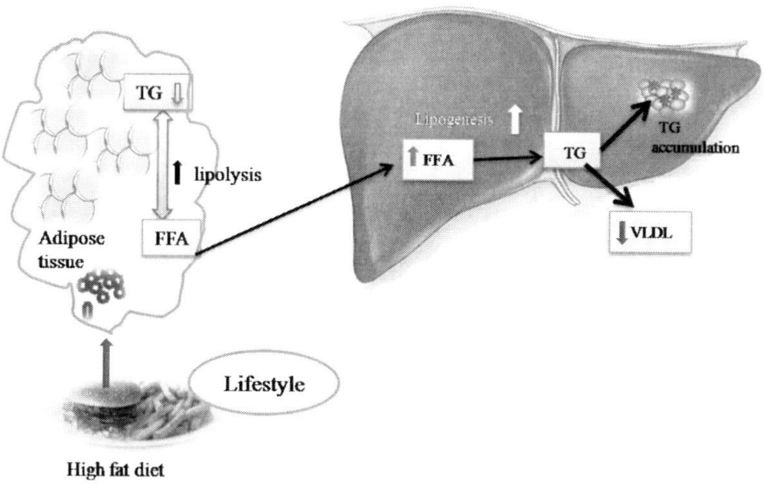

Figure 2. Metabolic mechanism involved in NAFLD: Consumption of high fat diet causes disturbance in fatty acid (FA) metabolism that result in increased free fatty acid (FFA) synthesis in the adipocytes. The normal site of FA accumulation is adipose tissue. When FFA is synthesized in excess, they are released into the blood plasma and reaches to liver where fatty acids are degraded to acyl CoA. In diseased condition, β-oxidation decreases and lipogenesis increased that enhance triglyceride synthesis (TG) which is normally converted to very low density lipoprotein but during NAFLD, it starts getting accumulated into hepatocytes in the form of fat droplets that initiate 1st stage of NAFLD i.e. steatosis.

This excludes other causes of liver disease, estimate prognosis, and determine progression of fibrosis. Although radiologic and serum chemistries have been suggested as alternative non-invasive methods for diagnosis. Insulin resistance should be estimated, as close relationship exists with NAFLD and metabolic syndrome. The Homeostasis Model Assessment (HOMA) is an easy method that provides an estimate of insulin resistance based on fasting serum glucose and serum insulin levels.

5.1. Radiological Imaging Methods

Ultrasonography, CT and MRI scanning are reliable tools to detect moderate to severe fatty changes in the liver (Saadeh et al., 2002). The sensitivity of each imaging method increases with degree of fatty acid infiltration; at least 33% steatosis is required for detection. Although ultrasound, CT scan, MRI, and proton magnetic spectroscopy have been used to diagnose NAFLD however, differentiation between NASH and steatosis is difficult with any of the radiologic modalities (Fishbein et al., 1997; McCullough et al., 2004).

5.2. Serum Biochemistry

Abnormal liver function test has been investigated in case of 75% NAFLD patients. Increased alanin aminotransferase activities are the most common abnormality in NASH patients (Clark et al., 2003; McCullough et al., 2004). Usually, ALT or AST level gets elevated from mild to moderate ranging from 2-5 folds. The AST/ALT ratio have been reported <1 in 65-90% of NAFLD patients. The AST/ALT ratio of >1 is suggestive of an advanced fibrotic form of NAFLD. The serum albumin and bilirubin levels are almost normal, unless cirrhosis develops and serum ferritin elevates in about 50% of patients without evidence of hepatic iron overload (Fargion et al., 2001).

5.3. Liver Biopsy

The "gold standard" for diagnosing NAFLD is clinic-pathological correlation, with confirmation of steatosis by liver biopsy. This is generally performed on patients who have persistently elevated liver enzyme without viral serologic markers of chronic liver disease. It is valuable tool for clinical confirmation to exclude other liver disease as well as for its monitoring of progression (Brunt et al., 1999). It also provides prognostic information by differentiating steatosis from NASH as well as determines stage of fibrosis. Since alcoholic liver disease and NAFLD have similar histologic features, hence they cannot be distinguished by means of liver biopsy. The decision to perform a biopsy should be made on case to case basis (Sorbi et al., 2000; Adams et al., 2005; Neuschwander et al., 2003).

6. Treatment and Management of Nafld

Since potential for liver disease progression occurs exclusively in patients with NASH, hence treatment strategies should be focused on its prevention. The disorder is associated with cardiovascular complications, it is thus important to treat components of this metabolic syndrome (hyperlipidaemia, hypertension and diabetes). At present, treatment options for NAFLD and NASH are limited such as gradual weight loss (10% over a period of 6 months) and regular exercise to improve visceral adiposity.

6.1. Weight Loss, Diet Restrictions and Physical Activities

Modest weight loss of 1 kg/week is associated with lower incidence of metabolic syndrome and improvement of histologic features of NASH in > 80% cases. Loss of 4-5% of body weight decreases aminotransferase and fasting insulin levels (Chin et al., 2008). This can be achieved by dietary changes alongwith or without exercise, or bariatric surgery. Several studies have indicated positive effects of dietary changes on NAFLD treatment

(Tendler et al. 2007; Baba et al. 2006; Thomas et al. 2006). One study showed that restricted diet of 25 calories /kg body weight and exercise regime over 3 months reduced liver enzyme levels and hepatic steatosis, however hepatic inflammation and fibrosis improvement status remained unknown. Exercise promotes increased muscle mass in addition to weight loss, thereby increase peripheral insulin sensitivity.

6.2. Pharmacotherapy

Various lipid-lowering drugs *viz.*, clofibrate, gemfibrozil, atorvastatin, pitavastatin and probucol have been tried in NASH treatment. Antidiabetic agents i.e. metformin, troglitazone, rosiglitazone and pioglitazone have been used to improve insulin sensitivity in patients with NASH. Metformin improves insulin sensitivity with anti- TNF action. However, till date no single pharmacological agents have been approved for treatment of NAFLD or NASH. Orlistat (an enteric lipase inhibitor) and sibutramine (serotonin and norepinephrine reuptake inhibitor) are approved medications for weight loss by US Food and Drug Administration (FDA). Sibutramine promotes satiety and increases energy expenditure by stimulating thermogenesis. Improvements in NAFLD patients treated with these medications mainly attributed to weight loss (Lam et al., 2010).

6.2.1. Insulin Sensitizing Drugs

Insulin resistance is a key player in development of NAFLD and NASH, hence, much attention has been paid on insulin sensitizers. Two classes of insulin sensitizers have been studied:

Biguanides mainly increase hepatic insulin sensitivity and reverse insulin resistance induced by tumor necrosis factor alpha.

Glitazones improve insulin sensitivity in both diabetic and euglycemic patients by activation of nuclear transcription factor called peroxisome proliferator-activated receptor gamma (PPARγ). These effects do not persist after treatment, and drugs are also associated with an average weight gain of 3-6 kg (Bugianesi et al., 2005)

Thiazolidinediones (TZDs) are peroxisomal proliferator activated receptor-γ (PPARγ) agonists that promote hepatic fatty acid oxidation, decrease hepatic lipogenesis, increase peripheral and hepatic insulin sensitivity. The effect of TZDs on inflammation, fibrosis, and other histological features of NASH are less uniform (Oh et al., 2008; Chalasani et al., 2009).

Metformin improves insulin resistance by reducing hepatic gluconeogenesis, lipogenesis and fatty acid oxidation, increase peripheral and hepatic insulin sensitivity, decrease intestinal glucose absorption, and lower serum lipid. Preliminary trials showed that metformin is relatively well tolerated in NASH patients, improve aminotransferases levels and insulin resistance (Bugianesi et al., 2005; Duseja et al., 2004).

6.2.2. Lipid-Lowering Drugs

Lipid-lowering drugs target the high levels of triglycerides and low levels of high-density lipoprotein cholesterol that often associated with insulin resistance and NAFLD. Statins and gemfibrozil (Lopid) lowers aminotransferase levels (Adams et al., 2005).

6.3. Complementary Medication for NAFLD

Complementary or alternative medicine (CAM) therapies are "medical and health care practices that are not an integral part of allopathic system of medicine." They can be classified into five categories: biological based therapies, mind-body medicine, alternative medicines, manipulative and body-based systems, and energy-based systems. Majority of people with chronic liver disease are interested in CAM remedies.

6.3.1. Herbs

According to World Health Organization, about two third of world population depends upon herbal remedies for health care. The uses of plant, their extracts or plant derived chemicals have become a therapeutic modality for disease management. A variety of herbs have been found useful in prevention, treatment, and management of chronic liver diseases patients. Herbs used in treatment of liver conditions are classified as cholagogues, which induce the liver to produce bile; choloretics, which increase flow of bile from gallbladder; and cholekinetics, stimulate bile duct contraction.

Milk Thistle for Chronic Liver Diseases

Milk thistle (*Silybum marianum*, Family; *Asteraceae*) extract, one of the most commonly used traditional therapies, contain number of flavonolignans produced in plants by radical coupling of a flavonoid and a phenylpropanoid. A mixture of these flavonolignans, termed silymarin, was first isolated from an extract of milk thistle fruit. Silymarin is known to composed of mainly silybin (~50–70%), silychristin, silydianin, and other closely related flavonolignans (Loguercio et al., 2011). Evidence suggests widespread use of silymarin compounds among patients diagnosed with chronic liver diseases. The effects of silymarin are due to antioxidant and hepatoprotective properties. Silymarin alters hepatocyte membrane, stimulate nucleolar polymerase A that increases protein synthesis, scavenge free-radical, and inhibit peroxidizing enzymes. Twenty hepatitis patients treated with silymarin-phosphatidylcholine complex for seven days showed decrease in lipid peroxidation and transaminases levels (Venkataramanan et al., 2000).

6.3.2. Lactic Acid Bacteria (LAB)

Lactic acid bacteria are gram-positive, nonsporing, nonrespiring cocci or rods, which produce lactic acid as the major end product during the fermentation of carbohydrates. Based on taxonomic revisions, 20 genera including lactic acid bacteria are important. However, with practical approach and food-technology point of view, following genera are considered principal LAB: *Aerococcus, Carnobacterium, Enterococcus, Lactobacillus, Lactococcus, Leuconostoc, Oenococcus, Pediococcus, Streptococcus, Tetragenococcus, Vagococcus, and Weissella.*

6.4. Probiotics Efficacy in NAFLD

Probiotics are mainly administered through functional foods such as dairy products. Lactobacilli and bifidobacteria are currently the most marketed probiotics worldwide. They exert multiple health effects, such as blocking of gastroenteric pathogens, neutralize food mutagens produced in the colon, enhance immune response, lower serum cholesterol, and control intestinal dysfunction (Chiu et al., 2006; Foo et al., 2003). In general, probiotics must colonize gastrointestinal tract (GIT) of the host, have acid and bile salt-tolerance, and inhibit putrefactive bacteria.

Although mechanisms of probiotic in NAFLD management are not fully elucidated, many effects such as modulation of intestinal microbiota, antibacterial substance production, epithelial barrier function, or immunomodulation may be involved. Gastrointestinal tract hosts >99% of microbial mass of the body. The intestinal microbiota is the principal source of pathogen-associated molecular patterns (PAMPs) both in health and disease. In addition to their role in innate immunity, toll like receptors (TLRs) also play a major role in regulation of inflammation. Probiotics can modulate intestinal micro-flora by competing for binding sites thus eliminating pathogens. They can also secret some antimicrobial components that inhibits growth of undesierable bacteria. Recent studies have demonstrated that beneficial effects were achieved not only by live bacteria but also by heat-inactivated or γ-irradiated non viable bacteria, isolated bacterial DNA or even probiotic-cultured media indicative of possible interaction between probiotic derived products, secondary metabolites or cell wall components with immune system (Stiles et al., 1997).

Table 2. Effect of probiotics on NAFLD related aspect in experimental models

Probiotics/dose (CFU/day)	Experiment model/ diet	Duration of therapy (weeks)	Inference	Reference
VSL#3 probiotic @ 1.5×10^9 CFU/ day	C57BL-6 ob/ob mice (n= 48)/ High fat diet*	4	Improves hepatic histology, serum alanine aminotransferase (ALT) elevations, and several biochemical measures of hepatic insulin resistance	Li et al., (2003)
Bacillus polyfermenticus SCD @ 3.1×10^6 CFU/day	Male Sprague Dawley rats (*n*_20)/ High-fat and high-cholesterol diet	6	Reduced plasma low-density-lipoprotein cholesterol, hepatic total cholesterol, and triglycerides, increased fecal excretion rates of total cholesterol and triglycerides	Paik et al.,(2005)
Lactobacillus rhamnosus PL60 @1.0×10^9 CFU/ day	Male C57BL/6J mice/ High fat diet*	8	Decreases body weight gain and liver steatosis in diet induced obese mice, anti-obesity effects of L. rhamnosus PL60	Lee et al., (2006)
VSL#3 probiotic @ 1.5×10^9 CFU/ day	Male C57BL/6J Mice/ High fat diet*	4	Improve HFD-induced hepatic NKT cell depletion, IR, hepatic steatosis and inflammation	Ma et al., (2008)
Lactobacillus plantarum MA2 1×10^{11} CFU/ day	Rats/ cholesterol-enriched diet	5	Reduce liver and serum cholesterol and triglycerides	Wang et al.,(2009)
VSL#3 1.3×10^{10} CFU/day	Rats/ High fat diet*	4	Amelioration of the hepatic inflammatory, steatotic and peroxidative factors and reduction in serum aminotransferase levels	Esposito et al., (2009)

(*High fat diet -that gives >60% energy from fat).

The mucus layer covering the gastrointestinal mucosa is first line of defense against mechanical, chemical or microbiological aggressions arising from the luminal contents. It has been demonstrated that lactobacilli up-regulate MUC2 and MUC3 mucins and inhibit attachment of enterohemorrhagic *Escherichia coli* (Kim et al., 2008). Probiotic mixture of lactobacilli and bifidobacteria increase the secretion of mucin, stimulating MUC2 gene expression in rat (Caballero et al., 2007). Administration of probiotics to improve gut health has gained significant momentum in recent.

In vivo studies showed that the administration of probiotics and/or prebiotics is effective in improvement of lipid profiles i.e. reduction of serum/plasma total cholesterol, LDL-cholesterol and triglycerides or increment of HDL-cholesterol. Anderson and Gilliland (1999) reported significant reduction in serum cholesterol by consumption of one daily serving of yogurt. They concluded that every 1% reduction in serum cholesterol concentration is associated with an estimated 2- 3% reduction in risk for coronary heart disease. Regular intake of fermented milks containing *L. acidophilus* has reduced risk for coronary heart disease by 6-10%. The cholesterol-reducing abilities was due to the assimilation of cholesterol by *L. acidophilus* cells or/and attachment of cholesterol to its surface. Bile salt deconjugation ability, bile salt hydrolase activity (BSH) of lactic acid bacteria and co-precipitation of cholesterol with deconjugated bile results fast removal and less absorption in gastrointestinal tract (Liong and Shah, 2005).

Administration of VSL#3 probiotic preparation improves liver histology and modulated intestinal microflora of obesity-related fatty liver disease (Li et al., 2003). An eight-week oral treatment with *L. rhamnosus* PL60 showed an antiobesity and liver steatosis effect in diet-induced obese mice. Histopathological analysis of liver steatosis evidenced low grading score in diet-induced obesity in mice receiving *L. rhamnosus. L. plantarum* MA2. Administration of these cultures in rats demonstrated reduction of both liver cholesterol and triglycerides, and increased lactobacilli and bifidobacteria population (Lee et al., 2006; Wang et al., 2009). In human trials, administration of probiotic VSL#3 improved LFTs, malondialdehyde, 4-hydroxynonenal, and S-nitrosothiol levels. Similarly, Solga et al.(2008) reported that steatosis improved in 3 out of 4 patients treated with VSL#3 for 4 months when measured by proton magnetic resonance spectroscopy. The above animal and human investigations have suggested beneficial effects of probiotics, however convincing evidence has yet to be demonstrated in humans, hence further studies are necessary to substantiate these results.

Conclusion

Globalization, electronic and print media as well as advancement in food processing technology have significantly contributed towards new life style changes. This has resulted in emergence of health related disorders. Among these, non alcoholic fatty liver disease has also become epidemic. There is lack of diagnostic tools to diagnose the disease in early stage and made its treatment very difficult in final stage. Present treatment options are also inadequate. The complementation or alternative treatment coupled with indigenous knowledge exhibits potential to overcome the menace of NAFLD. Among these, herbes or their bioactive components, probiotic lactic bacteria or combination thereof possesses treasure for effective treatment in synergistic way. However, before these recommendations, the extensive research on bioactive components, their effective doses, safety and delivery system is required to be established by precisely designed *in-vitro* and *in-vivo* experiments.

References

Adams LA, Angulo P, Lindor KD. Nonalcoholic Fatty Liver disease. *CMAJ,* 2005;172(7):899-905.

Amarapurkar D, Kamani P, Patel N. Prevalence of nonalcoholic fatty liver disease: population based study. *Ann. Hepatol.,* 2007;6:161-163.

Anderson JW and Gilliland SE. Effect of fermented milk (yogurt) containing *Lactobacillus acidophilus* L1 on serum cholesterol in hypercholesterolemic humans. *Journal of the American College of Nutrition,* 1999;18(1):43-50.

Angulo P, Keach JC, Batts KP, Lindor KD. Independent predictors of liver fibrosis in patients with nonalcoholic steatohepatitis. *Hepatology*, 1999;30:1356-1362.

Arun J, Clements RH, Lazenby AJ, Leeth RR, Abrams GA. The prevalence of nonalcoholic steatohepatitis is greater in morbidly obese men compared to women. *Obes. Surg.,* 2006;16:1351-1358.

Baba CS, Alexander GA, Kalyani B, Pandy R, Rastogi S, Pandey A. Effect of exercise and dietary modification on serum aminotransferases levels in patients with nonalcoholic steatohepatitis. *J. Gastroenterol. Hepatol.,* 2006;21:191-198.

Bacon BR, Farahvash MJ, Janney CG, Neuschwander-Tetri BA. Nonalcoholic steatohepatitis: *an expanded clinical entity. Gastroenterology,* 1994;107:1103-1109.

Bala S, Marcos M, Szabo G. Emerging role of microRNAs in liver diseases. *World J. Gastroenterol.,* 2009;15:5633-5640.

Bellentani S, Bedogni G, Miglioli L, Tiribelli C. The epidemiology of fatty liver. *Eur. J. Gastroenterol. Hepatol.,* 2004;16:1087-1093.

Bellentani S, Marino M. Epidemiology and natural history of nonalcoholic fatty liver disease. *Annals of Hepatology,* 2009;8(1):S4-S8.

Bellentani S, Saccoccio G, Masutti F, Croce L.S, Brandi G, Sasso F, Cristanini G, and Tiribelli C. Prevalence of and risk factors for hepatic steatosis in Northern Italy. *Ann. Intern. Med.,* 2000;132:112-117.

Brunt EM, Janney CG, Bisceglie AMD, Neuschwander TBA, Bacon BR. Nonalcoholic steatohepatitis: a proposal for grading and staging the histological lesions. *Am. J. Gastroenterol.,* 1999;94:2467-2474.

Bugianesi E, Gentilcore E, Manini R. A randomized controlled trial of metformin versus vitamin E or prescriptive diet in nonalcoholic fatty liver disease. *Am. J. Gastroenterol.,* 2005;100:1082-1090.

Caballero-Franco C, Keller K, De Simone C, Chadee K. The VSL#3 probiotic formula induces mucin gene expression and secretion in colonic epithelial cells. *Am. J. Physiol. Gastrointest Liver Physiol.,* 2007;292(1):G315-322.

Chalasani N.P, Sanyal A.J, Kowdley K.V, Robuck P.R, Hoofnagle J, Kleiner D.E. Pioglitazone versus vitamin E versus placebo for the treatment of non-diabetic patients with nonalcoholic steatohepatitis: Pivens trial design. *Contemp. Clin. Trials*, 2009;30:88-96.

Chang BH, Li L, Paul A, Taniguchi S, Nannegari V, Heird WC and Chan L: Protection against fatty liver but normal adipogenesis in mice lacking adipose differentiation-related protein. *Mol. Cell Biol.,* 2006;26:1063-1076.

Charlton M, Sreekumar R, Rasmussen D, Lindor K, Nair KS. Apolipoprotein synthesis in nonalcoholic steatohepatitis. *Hepatology,* 2002;35:898-904.

Chin HK, Younossi ZM. Nonalcoholic Fatty Liver disease: a manifestation of metabolic syndrome. *Cleveland Clinic Journal of Medicine,* 2008;75(10):721-728.

Chiu CH, Lu TY, Tseng YY, Pan TM. The effects of *Lactobacillus*-fermented milk on lipid metabolism in hamsters fed on high-cholesterol diet. *Appl. Microbiol. Biotechnol.,* 2006;71:238-245.

Clark JM, Brancati FL, Diehl AM. The prevalence and etiology of elevated aminotransferase levels in the United States. *Am. J. Gastroenterol.,* 2003;98:960-967.

Clark JM, Diehl AM. Defining nonalcoholic fatty liver disease: implications for epidemiologic studies. *Gastroenterology,* 2003;24:248-250.

Clouston AD, Powell ED. Nonalcoholic fatty liver disease: is all fat bad. *Internal Medicine Journal,* 2004;34:187-191.

Contos MJ, Cales W, Sterling RK, Luketic VA, Shiffman ML, Mills AS. Development of nonalcoholic fatty liver disease after liver transplantation for cryptogenic cirrhosis. *Liver Transpl.,* 2001;7:363-373.

Duseja A, Murlidharan R, Bhansali A, Sharma S, Das A, Das R. Assessment of insulin resistance and effect of metformin in nonalcoholic steatohepatitis-a preliminary report. *Indian J. Gastroenterol.,* 2004;23:12-15.

Esposito E, Iacono A, Bianco G, Autore G, Cuzzocrea S, Vajro P. Probiotics reduce the inflammatory response induced by a high-fat diet in the liver of young rats. *J. Nutr.,* 2009;139(5):905-911.

Fishbein MH, Gardner KG, Potter CJ, Schmalbrock P, Smith MA. Introduction of fast MR imaging in the assessment of hepatic steatosis. *Magn. Reson. Imaging,* 1997;15:287- 293.

Foo HL, Loh TC, Law FL, Lim YZ, Kufli CN, Rusul G. Effects of feeding *Lactobacillus plantarum* I-UL4 isolated from Malaysian tempeh on growth performance, faecal flora and lactic acid bacteria and plasma cholesterol concentrations in postweaning rats. *Food Sci. Biotechnol.,* 2003;12:403-408.

Fujita K, Nozaki Y, Wada K. Dysfunctional very-low-density lipoprotein synthesis and release is a key factor in nonalcoholic steatohepatitis pathogenesis. *Hepatology,* 2009;50:772-780.

Gokulakrishnan K, Anjana RM, Indulekha K, Anuradha S, Mohan V. Association of hypoadiponectinemia with non-alcoholic fatty liver disease in urban south Indians - (CURES - 81). *Indian J. Med. Res.,* 2010; 132:271-277.

Greco D, Kotronen A, Westerbacka J, Puig O, Arkkila P, Kiviluoto T, Laitinen S. Gene expression in human NAFLD, *Am. J. Physiol. Gastrointest Liver Physiol.,* 2008;294:G1281-G1287.

Grill JP, Cayuela C, Antoine JM, Schneider F. Effects of *Lactobacillus amylovorus* and *Bifidobacterium breve* on cholesterol. *Lett. Appl. Microbio*l., 2000;31:154–156.

Gupte P, Amarapurkar D, Agal S. Non alcoholic steatohepatitis in type 2 diabetes mellitus. *J. Gastroenterol. Hepatol.,* 2004;19:854-858.

Hand NJ, Master ZR, Le Lay J, Friedman JR. Hepatic function is preserved in the absence of mature microRNAs. *Hepatology,* 2009;49:618-626.

Herruzo JAS, Ruiz IG, Carreras MP, Yagüe MTM. Nonalcoholic fatty liver disease: from insulin resistance to mitochondrial dysfunction. *Revista Espanola De Enfermedades Digestivas,* 2006;98(11):844-874.

Iaconoa A, Rasoa GM, Cananib RB, Calignanoa A, MeliaIacono R. Probiotics as an emerging therapeutic strategy to treat NAFLD: focus on molecular and biochemical mechanisms, *Journal of Nutritional Biochemistry,* 2011;22(8):699-711.

Ip E, Farrell GC, Robertson G. Central role of PPAR alpha-dependent hepatic lipid turnover in dietary steatohepatitis in mice. *Hepatology,* 2003;38:123-132.

Ipekci SH, Basaranoglu M, Sonsuz A. The fluctuation of serum levels of aminotransferase in patients with nonalcoholic steatohepatitis. *J. Clin. Gastroenterol.,* 2003;36:371.

Itoh S, Yougel T, Kawagoe K. Comparison between nonalcoholic steatohepatitis and alcoholic hepatitis. *Am. J. Gastroenterol.,* 1987;82:650-654.

Jimba S, Nakagami T, Takahashi M, Wakamatsu T, Hirota Y, Iwamoto Y, Wasada T. Prevalence of nonalcoholic fatty liver disease and its association with impaired glucose metabolism in Japanese adults. *Diabet Med.,* 2005;22(9):1141-1145.

Jin X, Ye YF, Chen SH. MicroRNA expression pattern in different stages of nonalcoholic fatty liver disease. *Dig. Liver Dis.,* 2009;41:289-297.

Kim SY, Lim HJ, Lee WK. Isolation of cholesterol-lowering lactic acid bacteria from human intestine for probiotic use *J. Vet. Sci.,* 2004;5(4):391–395

Kim Y, Kim SH, Whang KY, Kim YJ, Oh S. Inhibition of *Escherichia coli* O157:H7 attachment by interactions between lactic acid bacteria and intestinal epithelial cells. *J. Microbiol. Biotechnol.,* 2008;18(7):1278-1285.

Kohjima M, Enjoji M, Higuchi N, Kato M, Kotoh K, Yoshimoto T, Tatsuya F, Yada M, Yada R, Harada N, Ryoichi T, Nakamuta M. Re-evaluation of fatty acid metabolism-related gene expression in nonalcoholic fatty liver disease. *International Journal of Molecular Medicine,* 2007;20:351-358.

Lam B, Younossi ZM. Treatment options for nonalcoholic fatty liver disease. *Ther. Adv. Gastroenterol.,* 2010;3(2):121-137.

Lee HY, Park JH, Seok SH, Baek MW, Kim DJ, Lee KE. Human originated bacteria, *Lactobacillus rhamnosus* PL60, produce conjugated linoleic acid and show anti-obesity effects in diet-induced obese mice. *Biochim. Biophys. Acta.,* 2006;1761(7):736-744.

Lettéron P, Sutton A, Mansouri A. Inhibition of microsomal triglyceride transfer protein: another mechanism for drug-induced steatosis in mice. *Hepatology,* 2003;38:133-140.

Li Z, Yang S, Lin H, Huang J, Watkins PA, Moser AB. Probiotics and antibodies to TNF inhibit inflammatory activity and improve nonalcoholic fatty liver disease. *Hepatology,* 2003;37(2):343-350.

Liong MT, Shah NP, Optimization of cholesterol removal by probiotics in the presence of prebiotics by using a response surface method. *Applied and Environmental Microbiology,* 2005;71(4):1745-1753.

Loguercio C, Festi D. Silybin and the liver: From basic research to clinical practice. *World J. Gastroenterol.,* 2011;17(18):2288-2301.

Ludwig J, Viggiano TR, McGill DB, Oh BJ. Nonalcoholic steatohepatitis: Mayo Clinic experiences with a hitherto unnamed disease. *Mayo Clin. Proc.,* 1980;55:434-438.

Ma X, Hua J, Li Z. Probiotics improve high fat diet-induced hepatic steatosis and insulin resistance by increasing hepatic NKT cells. J Hepatol. 2008;49(5):821-830.

Marchesini G, Bugianesi E, Forlani G, Cerrelli F, Lenzi M, Manini R. Nonalcoholic fatty liver, steatohepatitis, and the metabolic syndrome. *Hepatology,* 2003;37:917-923.

McCullough AJ. The clinical features, diagnosis and natural history of nonalcoholic fatty liver disease. *Clin. Liver Dis.,* 2004;8:521–533.

Fargion S, Mattioli M, Francanzani AL, Sampietro M, Tavazzi D, Fociani P. Hyperferritinemia, iron overload, and multiple metabolic alterations identify patients at risk for nonalcoholic steatohepatitis. *Am. J. Gastroenterol.,* 2001;96:2448-2455.

McGarry JD, Mannaerts GP and Foster DW, A possible role for malonyl-CoA in the regulation of hepatic fatty acid oxidation and ketogenesis. *J. Clin. Invest.,* 1977;60:265-270.

Mohan V, Farooq S, Deepa M, Ravikumar R, Pitchumoni CS. Prevalence of non-alcoholic fatty liver disease in urban south Indians in relation to different grades of glucose intolerance and metabolic syndrome. *Diabetes Res. Clin. Pract.,* 2009;84:84-91.

Musso G, Gambino R, De Michieli F. Dietary habits and their relations to insulin resistance and postprandial lipemia in nonalcoholic steatohepatitis. *Hepatology,* 2003;37:909-916.

Neuschwander-Tetri BA, Brunt EM, Wehmeier KR, Oliver D, Bacon BR. Improved nonalcoholic steatohepatitis after 48 weeks of treatment with the PPAR-gamma ligand rosiglitazone. *Hepatology,* 2003;38:1008-1017.

Neuschwander-Tetri BA, Caldwell SH. Nonalcoholic steatohepatitis: summary of an AASLD single topic conference. *Hepatology,* 2003;37:1202-1209.

Nobili V, Marcellini M, Devito R, Ciampalini P, Piemonte F, Comparcola D, Sartorelli MR, and Angulo P. NAFLD in Children: a prospective clinical pathological study an effect of lifestyle advice. *Hepatology,* 2006;44(2):458-465.

Oh MK, Winn J, Poordad F. Review article: diagnosis and treatment of non-alcoholic fatty liver disease. *Ailment. Pharmacol. Ther.,* 2008;28:503-522.

Omagari H, Kadokawa Y, Masuda JI, Egawa I, Sawa T, Hazama H. Fatty liver in non overweight Japanase adult: incidences and clinical characteristics. *J. Gastroenterol. Hepatol.,* 2002;17:1098-1105.

Pagano C, Soardo G, Esposito W, Fallo F, Basan L. Plasma adiponectin is decreased in nonalcoholic fatty liver disease. *European Journal of Endocrinology,* 2005;152:113-118.

Paik HD, Park JS, Park E. Effects of *Bacillus polyfermenticus* SCD on lipid and antioxidant metabolisms in rats fed a high-fat and high-cholesterol diet. *Biol. Pharm. Bull.,* 2005;28(7):1270-1274.

Pinto HC, Baptista A, Camilo ME, Valente A, Saragoca A, de Moura MC. Nonalcoholic steatohepatitis: Clinicopathological comparison with alcoholic hepatitis in ambulatory and hospitalized patients. *Dig. Dis. Sci.,* 1996;41:172-179.

Ruhl CE, Everhart JE. Determinants of the association of overweight with elevated serum alanine aminotransferase activity in the United States. *Gastroenterology,* 2003;124:71-79.

Saadeh S, Younossi ZM, Remer EM, Gramlich T, Ong JP, Hurley M. The utility of radiological imaging in nonalcoholic fatty liver disease. *Gastroenterology,* 2002;123:745-750.

Savage DB, Petersen KF, Shulman GI. Mechanisms of insulin resistance in humans and possible links with inflammation. *Hypertension,* 2005;45:828-833.

Shen L, Fan JG, Shao Y, Zeng MD, Wang JR, Luo GH. Prevalence of nonalcoholic fatty liver among administrative officers in Shanghai: an epidemiological survey. *World J. Gastroenterol.*, 2003;9:1106-1110.

Solga SF, Buckley G, Clark JM, Horska A, Diehl AM. The effect of a probiotic on hepatic steatosis. *J. Clin. Gastroenterol.*, 2008;42:1117-1119.

Sorbi D, Boynton J, Lindor KD. The ratio of aspartate aminotransferase to alanine aminotransferase: potential value in differentiating nonalcoholic steatohepatitis from alcoholic liver disease. *Am. J. Gastroenterol.*, 1999;94:1018-1122.

Sorbi D, McGill DB, Thistle JL, Therneau TM, Henry J, Lindor KD. An assessment of the role of liver biopsies in asymptomatic patients. *Am. J. Gastroenterol.*, 2000;95(11):3206-3210.

Stiles ME and Holzapfel WH. Lactic acid bacteria of foods and their current taxonomy. *Int. J. Food Microbiol.*, 1997;36(1):1-29.

Tarantino G, Conca P, Basile V, Gentile A, Capone D, Polichetti G, Leo EA. Prospective study of acute drug-induced liver injury in patients suffering from nonalcoholic fatty liver disease. *Hepatol. Res.*, 2007;37:410-415.

Thomas EL, Brynes AE, Hamilton G, Patel N, Spong A, Goldin RD. Effect of nutritional counseling on hepatic, muscle and adipose tissue fat content and distribution in nonalcoholic fatty liver disease. *World. J. Gastroenterol.*, 2006;12:5813-5819.

Venkataramanan R. Milk thistle, a herbal supplement, decreases the activity of cyp3a4 and and uridine diphosphoglucuronosyl transferase in human hepatocyte cultures. *Drug Metabolism and Disposition,* 2000; 28(11):1270–1273.

Villanueva CJ, Monetti M, Shih M. Specific role for acyl CoA:Diacylglycerol acyltransferase 1 (DGAT1) in hepatic steatosis due to exogenous fatty acids. *Hepatology,* 2009;50:434-442.

Wang Y, Xu N, Xi A, Ahmed Z, Zhang B, Bai X. Effects of *Lactobacillus plantarum* MA2 isolated from Tibet kefir on lipid metabolism and intestinal microflora of rats fed on high-cholesterol diet. *Appl. Microbiol. Biotechnol.*, 2009;84(2):341-347.

Wanles IR, Lentz JS. Fatty liver hepatitis (steatohepatitis) and obesity: an autopsy study with analysis of risk factors. *Hepatology,* 1990;12:1106-1110.

Xu A, Wang Y, Keshaw H. The fat-derived hormone adiponectin alleviates alcoholic and nonalcoholic fatty liver diseases in mice. *J. Clin. Invest.*, 2003;112:91-100.

In: Molecular Biology of Bacteria
Editors: R. Kumar Gaur and H. K. Gautam
ISBN: 978-1-62618-189-2
© 2013 Nova Science Publishers, Inc.

Chapter 9

Understanding the Mysterious Radioresistance and Genomic Integrity of *Deinococcus radiodurans* against Gamma Radiation: Implications in Radiation Response Modulation in Higher Organisms

Raj Kumar[1]* and Hemant K. Gautam[2]
[1]Radiation Biotechnology Group, Radiation Biosciences Division,
Institute of Nuclear Medicine and Allied Sciences, Delhi, India
[2]Institute of Genomics and Integrated Biology, Mall Road Delhi, India

Abstract

Extreme radioresistnace of *Deinococcus radiodurans* is remained a mystery to the scientific community. Despite, extensive research by several laboratories world wide, absolute mechanism adopted by *D. radiodurans* is not clear even after five decade of its discovery. *D. radiodurans* showed close phylogenetic homology with the thermophilic *Thermus* sp., which is known for their radiosensitive properties. Several mechanisms have been proposed to explain extreme radioresistance of *D. radiodurans* however, due to contradictory evidences, none of them is accepted unanimously. *D. radiodurans* showed excellent DNA repair within 12-24 hrs after irradiation whereas, no evidence is available to justify its extraordinary DNA repair capabilities in terms of superior DNA repair enzymes pool. Whole genome sequence analysis of *D. radiodurans* has been completed. To understand the proteomes of *D. radiodurans* under normal and irradiated conditions several studies have been carried out. Some workers proposed that *D. radiodurans* have multiple copy of genomes and therefore adopt recombinational route to repair the damaged DNA. However, several other radiosensitive bacteria have been discovered carrying multiple copies of genomes. Proteomic analysis demonstrated that *D.*

* Corresponding Author: Email: rajkumar790@yahoo.com.

radiodurans over expressed some protein such RecA, LexA, Mut Y, Mut T, Mut M etc. under irradiation, may contribute to its extreme radioresistance. However, conversely, homologs of these proteins has also been identified in other radiosensitive bacteria further dilutes the claim. Now, new concept i.e. proteins rather DNA is the main target of radiation damage was proposed. According to the new concept, *D. radiodurans* reserve some vital proteins capable to provide excellent protection to the oxidized DNA and other cellular components. Apart from that *D. radiodurans* may also have a reserve of some novel secondary metabolites and perhaps use them during oxidative stress to combat oxidative stress. Therefore, logically, novel cytoprotective proteins/ other metabolites of radioresistant bacteria can modulate radioresponse in higher organism. Future research in this direction may explore the possibility of radioprotective drug development for human application. However, we have to run a long way before reaching the final destination i.e. radioprotection to biological systems.

Keywords: *Deinococcus radiodurans,* radioresistance, recombination, DNA damage and repair, oxidative stress, gamma radiation, radioprotection

1. Introduction

Deinococcus radiodurans was initially discovered by Arthur W. Anderson at the Oregon Agricultural Experiment Station in Corvallis, Oregon in 1956 (Anderson et al., 1956). It was initially identify as radioresistant bacterium, but latter on found that it can survive in cold, dehydration, vacuum and acidic environment. The name *Deinococcus radiodurans* taken from the ancient Greek word, meaning of “terrible grain/berry" and the Latin word *radius* and *durare*, meaning "radiation surviving" (Huyghe, 1998). Primarily, *Deinococcus radiodurans* was classified and placed in the genus *Micrococcus*. However, letter on based on the ribosomal RNA gene sequencing and homology analysis, it was placed in the separate genus *Deinococcus*, closely related to the genus *Thermus*, the heat-resistant bacteria (Makarova et al., 2001). All other known members of the genus *Deinococcus* are *D. geothermal, (D. murray, D. proteolyticus, D. indicus, D. radiopugnans, D. frigens, D. radiophilus D. saxicola, D. grandis D. marmoris* and *D. deserti* (de Groot et al., 2005; Cox et al., 2005). However, unlike to other members, *D. radiodurans* is relatively large, tetrad spherical bacterium with a diameter of 1.5 to 3.5 μm (Makarova et al., 2001). *D. radiodurans* is non endospores forming and nonmotile bacterium. Physiologically, it is an obligate aerobic chemo-organo-heterotroph, i.e., uses oxygen to generate its energy from organic substances. *D. radiodurans* genome consists of two circular chromosomes, (Chromosome I 2.65 million bp; Chromosome II 412,000 base pairs). Apart from that it also has a mega (177,000 bp) and a small (46,000 bp) plasmids. *D. radiodurans* has about 3,195 genes. In stationary phase, *D. radiodurans* contains four copies of genome. While, in multiplying stage, it contains 8-10 copies of the genome. *D. radiodurans* withstand happily with no loss of viability up to 5,000 Gy of ionizing radiation. Whereas, at 15,000 Gy of gamma radiation dose, *D. radiodurans* showed ~37% survival (Moseley and Mattingly, 1971; Ito et al., 1983 and Murray, 1992). For comparison, a dose of 5,000 Gy is estimated to induce several hundred double-strand breaks (DSBs) in the DNA of an organism (~0.005 DSB/Gy/Mbp; haploid genome). 5 Gy dose of radiation is enough to kill a human within 30 days, while, 200-800 Gy can kill *E. coli*, and over 4,000 Gy can kill the radiation-resistant tardigrade. Several species of bacteria with comparable radioresistance have also been reported. Some of them include genus *Chroococcidiopsis* (phylum cyanobacteria) and some species of *Rubrobacter* (phylum actinobacteria). Among the Archaea, the bacterial species *Thermococcus gammatolerans* showed comparable radioresistance (Cox et al., 2005).

Natural background radiation level estimated as an average of 0.4 mGy/ year globally. The highest known background radiation level, i.e. near Ramsar, in Iran is only 260 mGy/year. In view of substantially low background radiation pressure, how a bacterium can acquire extreme radiation resistance? is an all time interesting but unreciprocated question before the world scientific community. Valerie Mattimore Battista has suggested that extreme radioresistance of *D. radiodurans* could be a results of a mechanism involved in the response to protection against environmental desiccation. It was observed that mutant strains of *D. radiodurans* which are susceptible to irradiation are also equally labile to damage from prolonged desiccation (Mattimore and Battista, 1996). Besides that, LEA proteins (Late Embryogenesis Abundant proteins) (Goyal et al., 2005), expressed by *D. radiodurans* to protect against desiccation was further provide a support to the hypothesis proposed by Battista et al., 2001). In an independent study, Michael Daly from Uniformed University of the Health sciences, Bethesda, USA has suggested that *D. radiodurans* accumulates manganese complexes and use it as antioxidants to protect itself against oxidative stress caused by radiation damage (Daly et al., 2007). Michael Daly and his team also concluded that high intracellular manganese (II) concentration in *D. radiodurans* provides protection to several proteins from radiation mediated oxidation. They also proposed the idea that "protein, rather than DNA, is the principal target of ionizing radiation in sensitive bacteria (Daly et al., 2007). Despite, the persisting ambiguities and unclear understanding about the remarkable mechanism of radioresistant used by *D. radiodurans,* various biotechnological applications of radioresistant bacteria has been explored. Therefore, to understand the mystery and secrecy of radioresistance mechanism of *D. radiodurans,* several studies were carried out in the past. Some of them are being discussed in subsequent sections of current review.

2. Comparative Evolutionary Genomics of Thermo/Radioresistant Bacteria

Deinococcus sp. and *Thermus* sp. are classified and kept in to entirely different branch of phylogeny under *Deinococcus-Thermus* (DT) group (Weisburg et al., 1989; Marina et al., 2005). The members of DT group seems to shear common origin that was supported by phylogenetic trees homology analysis of ribosomal proteins, RNA polymerase subunit, Rec A etc., (Weisburg et al., 1989; Rainey et al., 1997 and Gupta, 1998). The *Thermus* genus consists of 14 thermophilic species, whereas, the *Deinococcus* genus have only 7 species. All of them are mesophilic in nature and known for their extreme resistance against gamma irradiation, desiccation and other DNA damaging chemical agents (Battista et al., 1999; Makarova et al., 2001). However, interestingly, two species i.e. *D. geothermalis* and *D. murrayi*, are considered to be thermophilic (Ferreira et al., 1997).

Several evolutionary factors could contribute to the genome differentiation of *Thermus* (TT) and *Deinococcus* (DR) to the divergence from the common ancestor. Some examples includes differential gene loss and gain, horizontal gene transfer (HGT), loss of the ancestral orthologous gene, Lineage-specific expansion of paralogous gene families, or acquisition of paralogs via HGT (Marina et al., 2005).

3. Gene Loss and Gain in Thermus and *Deinococcus* sp

Reconstruction of the evolutionary events that occurred after the deviation of the *Thermus* and *Deinococcus* lineages from the common antecedent, several genes have been delineated. The absence of certain metabolic genes in *Thermus* group creates gaps in its metabolic pathways. However, *Thermus* group is able to biosynthesed all amino acids, nucleotides and majority of cofactors, that suggested that the gaps are crammed by analog enzymes. For example, *Thermus* and *Deinococcus* group encode unrelated gene for thymidylate synthase. Similarly, the classical, folate-dependent thymidylate synthase present in *Deinococcus radiodurans* is probably ancestral in bacteria. Folate-dependent thymidylate synthase was displaced via horizontal gene transfer (HGT) in the thermus lineage by the flavin dependent thymidylate synthetase, which is a typical charaecteristics of archea and other thermophilic bacteria. In other cases, the substituted analogous pathways are still remaining uncharacterized. In contrast, to *Deinococcus radiodurans* (DR), the genes that were acquired by thermos (TT), represent a clear correlation to the thermophilic way of life. In particular, TT seems to acquired 23 gene families from the set of putative thermophilic determinants (Makarova, 2003), while, *Deinococcus radiodurans* acquired only one. The preponderance of these proteins (17 of 31) are encoded in the TT megaplasmid while 11 belong to the predicted DNA repair system characteristic of thermophiles (Makarova et al., 2002; Jansen et al., 2002). In addition TT has acquired 4 "archeal" genes that are not encoded in any of the genomes of mesophilic bacteria assigned to COGs (Peptide chain release factor 1, DNA modification methylase, and two membrane protein, COG3462 and COG4645).

Majority of genes lost by *D. radiodurans* during evolution encode enzymes of energy metabolism and biosynthesis of some critical cofactors. For example, the loss of the three subunits of pyruvate: ferredoxin oxidoreductase, a key reaction of central metabolism. Another example of gene loss in *D. radiodurans* includes the three subunits of NAD/NADP transhydrogenase system, which is conscientious for energy dependent reduction of NAD+ (Sauer et al., 2004). Further, *D. radiodurans* lineage lost four enzymes of NAD biosynthesis and six enzymes of cobalamine biosynthesis. Resulted *D. radiodurans* is dependent on an exogeneous source of NAD for growth. In continuation, other genes acquired by the *D. radiodurans* lineage encode systems of protein degradation and amino acid catabolism (examples like, urease, urea transporter, hidtidine degradation system, monoamine oxidase, lysine 2, 3-aminomutase, kynureninase and tryptophan 2-3 dioxygenase, peptidase, and corboxy-peptidase etc.). In view of above, it can be concluded that acquisition and expansion of metabolic systems, on the cost of certain biosynthetic capabilities, support the hypothesis of metabolic restructuring that could impact the resistance against oxidative stress by decreasing the energy requirements for cellular activities (Imlay, 2003).

4. Highly Expressed Genes of *Deinococcus radiodurans* under Radiation Pressure

To address the queries i. does *D. radiodurans* possess novel repair proteins that are more effective than the known proteins in other species? ii. Does *D. radiodurans* use common repair machinery more efficiently? Predicted highly expressed (PHX) and putative alien genes analysis was carried out (Karlin and Mrazek, 2001). The outcome of the study indicated that DR possesses a collection of PHX S-layer (surface structure) proteins that may provide environmental protection. Surprisingly, many ABC transporter genes (i.e. peptide, branched chain amino acids, Phosphates and various sugars) are PHX in nature. Aconitase also act as oxygen radical sensor (Hentze and Kuhn, 1996; Rouault, 1996). Aconitase (E(g) value 2.56) apparently alerts the cell to large amounts of oxidative damage. Aconitase has an active iron sulfur (4Fe-4S) core at the catalytic site and also regulates gene expression of the iron transport receptor and iron storage ferretin. Aconitase PHX value in other genomes is less compared to *D. radiodurans* (2.56) for example in *E coli* 1.4., *B subtilis* (1.08), M tuberculosis (1.18) and *A. aeolicus* (1.24).

Major PHX gene clusters include an operon of ribosomal proteins, with the EF-G (fus-1) and EF-Tu (Tuf-1) gene interspersed extending about 12kb on chromosome I. A second PHX cluster consists of the V-type ATP synthetase genes subunits I, K, E, A, B and D in the region 710-719 kb. A third PHX cluster involve several TCA and electron transposrt genes covering 959-967kb. The subsequent cluster contains four contiguous peptide ABC transporters covering 1585-1592 kb (Karlin and Mrazek, 2001). There is a PHX cluster of three ORF DR 2179, DR 2180 and DR 2181 proximal to position 2178 kb. A PA cluster on chromosom II features seven genes of lipopolysaccharide biosynthesis. The number of PHX proteins of unknown function is strikingly high. Many of these might be novel proteins contributing towards radiation resistance in *Deinococcus radiodurans*.

5. Mn Accumulation Contributes to Radioresistance in *Deinococcus radiodurans*

It has been reported that *Deinococcus radiodurans* accumulates manganese in large amounts during irradiation (Chou and Tan, 1990; Leibowitz et al., 1976). Mn (II) was reported to be needed for DNA repair enzymes activity in *Deinococcus radiodurans* (Evan and Moseley, 1985) and thus implicated a significant role in its high radioresistance (Wierowski and Bruce, 1980). Mn (II) has been reported to have privileged binding with the guanine moiety in nucleotides, dinucleotides monophosphate and DNA of cells having high G+C contents and altered interaction between DNA and proteins (Eisinger et al., 1965; Anderson et al., 1971; Murray, 1986).

The growth of *Deinococcus radiodurans* in DMM medium was reported to be dependent on Mn (II) but not on Fe, Co, or Mo etc. Moreover, it also been reported that growth and radioresistance of *D. radiodurans* was unaffected by Fe chelators (Daly et al., 2004). Therefore, it is hypothesized that Mn(II) play a critical role in *D. radiodurans* survival against ionizing radiation that is independent of Mn-SOD. The existence of common trait i.e. high interacellular Mn/Fe concentration ratio in phylogenetically separate radioresistant bacteria compared to sensitive bacteria, supports the idea that Mn (II) accumulation (with low Fe) might be a widespread tactic that contributes to radioresistance.

6. Efficient Repair of Oxidized DNA Bases in *Deinococcus radiodurans*

To protect against oxidative DNA damage caused by ionizing radiation, cells have adapted efficient and accurate repair mechanism to remove oxidize lesions from DNA (Demble and Harrison, 1994; Bauche and Laval, 1999). In *E. coli,* three DNA glycosylases are reported to recognize oxidized purine and pyrimidines in DNA.

Similarly, Fpg protein (formamidopyrimidine-DNA glycosylase) recognizes oxidized purines such as 2,6-diamino-4-hydro-8-oxoguanine (8-OxoG) residues and get rid of them from DNA. 8-OxoG is the main oxidative lesion produced in DNA by reactive oxygen species (Grollman and Moriya, 1993) and carried mutagenic trait as reported *in vivo* and *in vitro* (Shibutani et al., 1991; Wood et al., 1994).

Similarly, noncoding oxidized pyrimidines (such as thymine glycol, cytosine glycol and N-substituted urea (Evans et al., 1993; Maccabee et al., 1994) are recognized and removed by Nth (Asahara et al., 1989; Demble and Harrison, 1994) and /or Nei (Jiang et al., 1997; Melamede et al., 1991) proteins.

These three glycosylases have an apurinic/apyrimidine (AP) lyase activity which cleaves the DNA backbone by a beta elimination mechanism. (Boiteux et al., 1990; Kim and Linn, 1988).

In addition, Fpg and Nei proteins also catalyase excision of 5' terminal deoxyribose phosphate from DNA and thus have DNA deoxyribose phosphodiesterase (dRpase) activity (Graves et al., 1992). *D. radiodurans* has four proteins associated with ionizing radiation resistance: i.e. RecA, Pol g irrB and irrI genes but there biochemical function yet to be determined (Udupa et al., 1994; Mattimore et al., 1995). To understand the extreme radioresistance system operating in *D. radiodurans*, several mutant *D. radiodurans* strains were developed by site directed targeted mutagenesis methods.

The results of the study demonstrated that *D. radiodurans* mutant strains unable to express catalase (KatA gene) and superoxide dismutase (SOD A gene) were found to be more radiosensitive compared to wild type *D. radiodurans*, further suggested that antioxidant enzyme play critical role to combat oxidative stress in *D. radiodurans*. It has also been demonstrated that *D. radiodurans* contains DNA repair enzymes i.e. Uracil-DNA glycosylase and an apurine/apyrimidinic (AP) endonuclease, could be an integral part of its DNA base excision repair system.

It was further demonstrated that DNA-deoxyribo-phosphodiesterase activity used to incised AP sites in DNA was also present in the cell lysate of *D. radiodurans* (Mun et al., 1994). However, interestingly, partially purified dRpase activity was found to be lower in *D. radiodurans* compared to *E. coli*.

7. Presence of Specific Proteins Bind with Z Form of DNA in *Deinococcus radiodurans*

Duplex DNA in solution can adopt various configurations including B, Z, H, X form of DNA etc. B and Z form might be important because they have opposite twist. It is known that insertion of Z-form segments into cellular DNA, stimulates genetic recombination at the B-Z junction. A protein which preferably binds with Z-form duplex DNA has been separated out from *D. radiodurans* (Kitayama et al., 1988). Amino acid analysis of the protein indicates that it is not so basic, since it contains a lower mole percent of lysine and higher mole percent of aspartic acid similar to histone like DNA biding proteins of *E. coli* (Kitayama et al., 1988). Presence of this novel Z-DNA binding protein may provide some insight in extreme radioresistant of *D. radiodurans* in future.

8. Interchromosomal Recombination and Radioresistance in *Deinococcus radiodurans*

Despite the fact that efficient chromosomal DNA repair is the prominent cause of extreme radioresistance of *D. radiodurans*. Sole mechanism of its DNA repair pathway (s) has not been understood completely. Followed by gamma irradiation (1.5Mrad), approximetly, 100 DSBs per chromosomes observed in *D. radiodurans,* which were found to repaired within 12 to 24h without bacterial lethality (Daly et al., 1994; Daly et al., 1995). *D. radiodurans* contains 4 chromosomes per cells during stationary phase and up to 10 chromosomes per cell during exponential phase. Therefore, "multiple genome copies" hypothesis was proposed to explain its extreme radioresistance (Hansen, 1987). However, polyploidy in prokaryotes is very common and does not contribute to extreme resistance to radiosensitive organisms. For example, *Micrococcus luteus* and *M. sodonensis*, which carry multiple copies of chromosomes per cells, but are radiosensitive (Moseley, 1983). Similarly, *Azotobacter vinelandii*, which carry 40-80 chromosomes (Sadoff et al., 1979; Punita et al., 1989) per cell, is extremely sensitive to UV radiation (Majumdar and Chandra, 1985). Thus it is unlikely to quote that number of chromosomal copies hold the key of radioresistant in *D. radiodurans*. Howevere, *D. radiodurans* may uses redundant information in a manner that other organisms do not. The mechanism of recombinational DNA repair in *D. radiodurans* in the initial few hours followed by gamma irradiation was found completely differs from that occurring over the next 24h. The initial repair was observed found to be RecA independent, since it occurs to the same extent in both MD424 $recA^+$ and MD431 $recA^-$ mutant strains of *D. radiodurans* (Daly and Minton, 1996).

To understand whether extra chromosomal DNA like plasmid is actually accessible to the main chromosome during repair, studies on recombination between homologous chromosome and plasmids followed by high dose irradiation was carried out (Daly and Minton, 1996). The outcome of the investigation was indicated that, plasmid-chromosome recombination followed by irradiation of *D. radiodurans,* is largely Rec-A dependent. It was clearly observed that homologous plasmid and chromosomal fragments produced by gamma irradiation are accessible to each other in wild type *D. radiodurans* and can cross over during

DNA repair. These results suggested that *D. radiodurans* treat plasmid and chromosome in the same way. However, reports are also indicated non-receiprocal recA-independent plasmid-chromosomal recombination in *D. radiodurans*. Therefore, by enlarge a clear cut understanding on recombinational repair in *D. radiodurans* has not been established.

9. Novel Genes Participate in DNA Repair in Deinococcus radiodurans

Apart from Rec A, several other DNA repair genes reported to participate during post irradiation recovery, are present in the *D. radiodurans* (Carroll et al., 1996). To protect genomic integrity against oxidative DNA damage caused by ionizing radiation, *D. radioduran* cells have evolve efficient repair mechanism to delete damage DNA lesion (Narumi et al., 2001; Hansen, 1980). 7, 8-dihydro-8-oxoguanine (GO) is the most stable oxidized product of irradiated DNA. More interestingly, if not repaired GO lesion in DNA can produced A/GO mismatches during DNA replication (Sun et al., 2995) and lead to transversions from G* C to T* A (Wood et al., 1990; Cheng et al., 1991; Moriya et al., 1991; Moriya, 1993). In *E. coli*, three enzymes, Mut Y, Mut M (Fpg) and Mut T are known to involve in defense against the mutagenic effects of GO lesion (Micheal and miller, 1992; Tchou et al., 1991). Homologs of putative MutY, Mut M and Mut T are present in *D. radiodurans* (Makarova et al., 2001). The kinetic parameters of Mut Y protein of D. *radiodurans* were found very similar to the Mut Y of *E. coli*. Possitively, Mut Y protein of *D. radiodurans* has similar substrate specificity and binding preference for an A/GO mismatch over an A/G mismatch, as Mut Y of *E. coli* does (Lu and Li, 2001). Therefore, in conclusion *D. radiodurans* Mut Y protein can protect the cell from GO mediated mutations caused by ionizing radiation and oxidative stress.

10. Future Implications of Radioresistance Bacteria in Modulation of Radioresponse in Higher Animals to Achieve Efficient Radioprotection

Despite the several attempts including molecular studies, whole genome sequencing, proteomic analysis and bioinformatic analysis etc., the exact mechanism of extreme radioresistance in *D. radiodurans* was not established till date. Several report indicated that *D. radiodurans* do not have extra ordinary DNA repair system. It has similar gene involve in DNA repair as *E. coli*. There is no substantial difference in the gene expression in *D. radiodurans* compared to *E. coli* related to oxidative stress mitigating and DNA repair supporting components. Now focus is being shifting on proteins having ability to protect against gamma radiation induced oxidative damage. The proteins involve in repair and also regulate other vital signaling pathways may logically contributes to radioresistance to *D. radiodurans*. Therefore, new understanding is being developing on the hypothesis that probably *D. radiodurans* do not have highly advance DNA repair efficiency but have extraordinary protein protecting capability. Better protein protection logically is an integral part of the radioresistance. During irradiation, the cellular biosynthesis or transcription remained on hold in all life form including bacteria and thus no new gene transcription and subsequent protein synthesis may occur. When we irradiate radioresistant bacteria with extreme gamma radiation doses (i.e. 20 kGy), it will require to place the culture in irradiation chamber for at least hour(s) depending on the dose rate of the irradiation source. Therefore, bacterium, whose transcription is on hold cannot synthesis new proteins during irradiation process. Thus, we propose a hypothesis that radioresistant bacteria may have reserve pool of some novel compounds/ secondary metabolites (Rakesh et al. 2011; Kumar et al. 2011a; Kumar et al. 2011b) or proteins which do not denatured by extremely high doses of radiation and could perform DNA and protein protection function not only during irradiation but also at post irradiation time (Kumar et al., 2010; Kumar and Singh, 2012). Thus identification and characterization of such novel proteins/ other secondary metabolites may be helpful to understand long standing puzzle of extreme radioresistance of bacteria. Most importantly, reserve protein/ secondary metabolites pools of radioresistant bacteria can be used to achieve radioprotection in higher animals and perhaps in human in future.

References

Anderson A W, Nordan H C, Cain R F, Parrish G and Duggan D (1956). Studies on a radio-resistant micrococcus. I. Isolation, morphology, cultural characteristics, and resistance to gamma radiation. *Food Technol.* 10 (1): 575–577.

Anderson J A, Kuntz G P P, Evans H H and Swift T J (1971). Preferential interaction of manganous ions with the guanine moiety in nucleosides, dinucleoside monophosphates, and deoxyribonucleic acid. *Biochem.* 10, 4368-4374.

Asahara H, Wistort P M, Bank J F, Bakerian R H and Cunningham R P (1989). Purification and characterization of *Escherichia coli* endonuclease III from the cloned nth gene. Biochem. 28, 4444-4449.

Battista J R, Earl A M, Park M J (1999). Why is *Deinococcus radiodurans* so resistant to ionizing radiation? *Trends Microbiol.* 7(9), 362-365.

Battista JR, Park MJ and McLemore AE (2001). Inactivation of two homologues of proteins presumed to be involved in the desiccation tolerance of plants sensitizes *Deinococcus radiodurans* R1 to desiccation. *Cryobiol.* 43 (2): 133–139.

Bauche C and Laval J (1999). Repair of oxidized bases in the extremely radiation resistant bacterium *Deinococcus radiodurans. J. Bacteriol.* 181 (1), 262-269.

Boiteux S, O'Connor T R, Lederer F, Gouyette A and Laval J (1990). Homogeneous *Eschericluia coli* FPG. A DNA glycosylase which excises imidozole ring-opened purines and nicks DNA at apurinic\apyrimidine sites. *J. Biol.Chem.* 265, 3916-3922.

Carroll DJ, Michael J D, and Minton KW (1996). Expression of recA in *Deinococcus radiodurans. J. Bacteriol.* 178 (1), 130-135.

Cheng KC, Cahill DS, Kasai H, Nishimura S and Loeb LA (1991). 8-hydroxyguanine, an abundant form of oxidative DNA damage, causes G-T and A-C substitutions. *J. Biol. Chem.* 267, 166-172.

Chou F I and Tan S (1990). Manganese (II) induces cell division and increases in superoxide dismutase and catalase activities in an aging Deinococcal culture. *J Bacteriol.* 172(4), 2029-2035.

Cox MM and Battista JR (2005). *Deinococcus radiodurans*-the consummate survivor. *Nature Rev. Microbiol.* 3 (11): 882–92.

Daly M J and Minton K W (1995). Interchromosomal recombination in the extremely radioresistant bacterium *Deinococcus radiodurans. J. Bacterial.* 177, 5495-5505.

Daly M J and Minton K W (1996). Recombination between resident plasmid and the chromosome following irradiation of the radioresistant bacterium *Deinococcus radiodurans. Gene.* 187, 225-229.

Daly M J, Gaidamakova E K, Matrosova V Y, vasilenko A, Zhai M, Venkateswaran A, Hess M, Omelchenko M V, Kostandarithes HM, Makarova K S, Wackett L P, Fredrickson J K and Ghosal D (2004). Accumulation of Mn (II) in *Deinococcus radiodurans* facilitates gamma radiation resistance. *Science.* 306, 1025-1028.

Daly M J, Ouyang L, Fachs P and Minton K W (1994). *In vivo* damage and recA-Dependent repair of plasmid and chromosomal DNA in the radiation-resistant bacterium *Deinococcus radiodurans. J. Bacteriol.* 176, 3508-3517.

Daly, M J, Elena K, Gaidamakova, V Y, Matrosova, Alexander V, Min Z, Richard D. Leapman, B L, Bruce R, Shu-Mei W. Li, K M. Kemner, J and Fredrickson K (2007). Protein oxidation implicated as the primary determinant of bacterial radioresistance. *PLoS Biol.* 5 (4): e92 EP.

de Groot A, Chapon V, Servant P, Christen R, Saux MF, Sommer S and Heulin T (2005). *Deinococcus deserti* sp. nov., a gamma-radiation-tolerant bacterium isolated from the Sahara Desert". *Int. J. Syst. Evol. Microbiol.* 55 (6): 2441–2446.

Demple B and Harrison L (1994). Repair of oxidative damage to DNA: Enzymology and biology. *Annu. Rev. Biochem.* 63, 915-948.

Eisinger J, Fawaz-Estrup F and Shulman R G (1965). Binding of Mn^{2+} to nucleic acids. *J.Chem. Phys.* 42, 43-53.

Evans D M and Moseley B E B (1985). Identification and initial characterisation of a pyrimidine dimer UV endonuclease (UV endonuclease ß) from *Deinococcus radiodurans*: a DNA repair enzyme that requires manganese ions. *Mutat. Res.* 145, 199-128.

Evans J M, Maccabee M, Batahet Z J, Courcelle R, Bockrath H I and Wallace S (1993). Thymine ring saturation products: Lesion bypass, misinsertion and implications for mutagenesis. *Mutant. Res.* 299, 147-156.

Ferreira A C, Nobre M F, Rainey F A, Silva M T, Wait R, Burghardt J, Chung A P, de-Costa M S (1997). *Deinococcus geothermalis* sp. nov. and *Deinococcus murrayi* sp. nov., two extremely radiation-resistant and slightly thermophilic species from hot springs. *Int. J. Syst. Bacteriol.* 47(4), 939-947.

Goyal K, Walton LJ and Tunnacliffe A (2005). LEA proteins prevent protein aggregation due to water stress. *Biochem. J.* 388 (1): 151–157.

Graves R J, Felzenswalb I, Laval J and O'Connor, T R (1992). Excision of the 5'-terminal deoxyribose phosphate from damaged DNA is catalyzed by Fpg protein of *Escherichia coli. J. Biol. Chem.* 267, 14429-14435.

Grollman A P and Moriya M (1993). Mutagenesis by 8-oxoguanine: an enemy within. *Trends Genet.* 9, 246-249.

Gupta RS (1998). Protein phylogenies and signature sequences: A reappraisal of evolutionary relationships among Archaebacteria, Eubacteria, and Eukaryotes. Microbiol. *Mol. Biol. Rev*. 62(4), 1435-1491.

Hansen M T (1987). Multiplickty of genome equivalents in the radiative-resistant bacterium *Micrococcus radiodurans. J. Bacteriol*. 134, 71-75.

Hansen MT (1980). Four protein synthesized in response to deoxyribonucleic acid damage in *Micrococcus radiodurans. J. Bacterial*. 141, 81-86.

Hentze M W and Kuhn L C (1996). Molecular control of vertebrate iron metabolism: mRNA based regulatory circuits operated by iron, nitric oxide and oxidative stress. *Proc. Natl. Acad. Sci*. 93, 8175-8182.

Huyghe P (1998). "Conan the Bacterium" Sciences. 16–19.

Imlay J A (2003): Pathways of oxidative damage. *Annu. Rev. Microbiol*. 57, 395-418.

Ito H, Watanabe H, Takeshia M and Iizuka H (1983). Isolation and identification of radiation-resistant cocci belonging to the genus *Deinococcus* from sewage sludges and animal feeds. *Agric. Biol. Chem*. 47: 1239–47.

Jansen R, Embden J D, Gaastra W and Schouls L M (2002). Identification of genes that are associated with DNA repeats in prokaryotes. *Mol. Microbiol*. 43(6), 1565-1575.

Jiang D, Hatahet A, Blaisdell J O, Melamede R J and Wallace S S (1997). *Escherichia coli* endonuclease VIII: cloning, sequencing, and over-pressing of the nei structural gene and characterization of nei and nei mis-mutants. *J. Bacteriol*. 179, 3773-3782.

Karlin A and Mrazek J (2001). Predicted highly expressed and putative alien genes of *Deinococcus radiodurans* and implications for resistance to ionizing radiation damage. *Proc. Natl. Acad. Sci*. 98(9), 5240-5245.

Kim J and Linn S (1988). The mechanisms of action of *E. coli* endonuclease III and T4 UV endonuclease (endonuclease V) at AP sites. *Nucleic acids Res*.16, 1135-1141.

Kitayama S, Matsumura O and Masuda S (1988). Islation of a DNA binding protein from Deinococcus radiodurans having an affinity for a Z-form polynucleotide. *J. Biochem*. 104, 127-130.

Kumar R, Patel DD, Bansal DD, Mishra S, Mohammed A, Arora A, Sharma A Sharma R K and Tripathi R P (2010): Extremophiles and their futuristic applications. Sustainable Biotechnology: Source of renewable energy. *Springer publication* USA 279-294.

Kumar R and Singh A (2012): Smart Therapeutics from Extremophiles: Unexplored Applications and Technological Challenges. Extremophiles Biotechnology, Springer publication USA (In Press)

Kumar R, Sharma R K, Bansal D D, Patel D D, Mishra S, Miteva L, Dobreva Z, Gadjeva V and Stanilova S (2011a). Induction of immunostimulatory cytokine genes expression in human PBMCs by a novel semiquinone glucoside derivative (SQGD) isolated from a *Bacillus* Sp. INM-1. *Cellul. Immunol*. 267, 67-75.

Kumar R, Bansal D D, Patel D D, Mishra S, Karamalakova Y, Zheleva A, Gadjeva V and Sharma R K (2011b). Antioxidative and radioprotective activities of semiquinone glucoside derivative (SQGD) isolated from *Bacillus* sp. INM-1. *Mol. Cellul. Biochem*. DOI 10. 1007/s11010-010-0660-x.

Leibowitz P J, Schwartzberg L S and Bruce AK (1976). The *in vivo* association of manganese with the chromosome of *Micrococcus radiodurans*. *Photochem. Photobiol*. 23, 45-50.

Levin-Zaidman S, Englander J, Shimoni E, Sharma AK, Minton KW and Minsky A (2003). Ring like structure of the *Deinococcus radiodurans* genome: a key to radioresistance?. *Science.* 299, 254–256.

Li X and Lu A L (2001). Molecular cloning and functional analysis of the Mut Y homolog of *Deinococcus radiodurans.* 183, (21), 6151-6158.

Maccabee M, Evans J S, Glackin M P, Hatahet Z, Wallace S (1994). Pyrimidine ring fragmentation products. Effects of lesion structure and sequence context on mutagenesis. *J. Mol. Biol.* 256, 514-530.

Majumdar S and Chandra A K (1985). UV repair and mutagenesis in *Azotobacter vinelandii.* Zentralbl. *Mikrobiol.* 140, 247-261.

Makarova K S L, Aravind Y I, Wolf R L, Tatusov K W, Minton E, Koonin V and Daly M J (2001). Genome of the extremely radiation-resistant bacterium *Deinococcus radiodurans* viewed from the perspective of comparative genomics. *Microbiol. Mol. Boil.Rev.,* 65 (1): 44–79.

Makarova K S, Aravind L, Grishin N V, Rogozin I B, Koonin E V (2002). A DNA repair system specific for thermophilic Archaea and bacteria predicted by genomic context analysis. *Nucleic Acids Res.* 30(2), 482-496.

Makarova K S, Aravind L, Wolf Y I, Tatusov R L, Minton K W, Koonin E V, Daly M J (2001). Genome of the extremely radiation-resistant bacterium *Deinococcus radiodurans* viewed from the perspective of comparative genomics. *Microbiol. Mol. Biol. Rev.* 65(1), 44-79.

Makarova K S, Wolf Y I and Koonin E V (2003). Potential genomic determinants of hyperthermophily. *Trends Genet.* 19(4),172-176.

Marina V O, Wolf Y I, Gaidamakova E K, Matrosova V Y, Vasilenko A, Zhai M, Daly M J, Konin E V, Makrova K S (2005). Comparative genomics of *Thermus Therophilus* and *Deinococcus radiodurans*: divergent routes of adaptation to thermophily and radiation resistance. *BMC Evolutionary Biol.* doi: 10.1186/1471-2148-5-57.

Mattimore V and Battista JR (1996). Radioresistance of Deinococcus radiodurans: functions necessary to survive ionizing radiation are also necessary to survive prolonged desiccation. *J. Bacteriol.* 178 (3): 633–637.

Mattimore V, Udupa K S, Berne G A and Battista J R (1995). Genetic characterization of forty ionizing radiation-sensitive strains of *Deinococcus radiodurans* : linkage information from transformation. *J. Bacteriol.*177, 5232-5237.

Melamede RJ, Hatahet Z, Kow Y W, Ide H and Wallace S S (1991). Isolation and characterization of endonuclease VII from *Escherichia coli. Biochem.* 33, 1255-1264.

Michaels M L and Miller J H (1992). The GO system protects organisms from the mutagenic effect of the spontaneous lesion 8-hydroxyguanine (7,8-dihydro-8-oxo-guanine). *J. Bacteriol.* 174, 6321-6325.

Mokorova K S, Aravind L, Wolf Y I, Tatusov R L, Minton K W, Koonin EV and Daly M J (2001). Genome of the extremely radiation resistant bacterium *Deinococcus radiodurans* viewed from the perspective of comparative genomics. *Microbiol. Mol. Rev.* 65, 44-79.

Moriya M (1993). Single stranded shuttle phagemid for mutagenesis studies in mammalian cells: 8-oxoguanine in DNA induces targeted G-C to T-A transverseions in Simian kidney cells. *Proc. Natl. Acad. Sci.* 90, 1122-1126.

Moriya M, Ou C, Bodepudi V, Johnson F, takeshita M and Grollman A P (1991). Site directed mutagenesis using a gapped duplex vector: a study of translesion synthesis past 8-Oxodeoxyguanine in *Escherichia coli*. *Mutant Res.* 254, 281-288.

Moseley B E B (1983). Photobiology and radiobiology of *Micrococcus (Deinococcus) radiodurans*. *Photochem. Photobiol. Rev.* 7, 223-274.

Moseley B E B and Mattingly A (1971). Repair of irradiated transforming deoxyribonu- cleic acid in wild type and a radiation- sensitive mutant of *Micrococcus radiodurans*. *J. Bacteriol.* 105 (3): 976–83

Mun C, Rowe, J D, Sandigursky M, Minton K W and Franklin W A (1994). DNA deoxyribophosphodiesterase and an activity that cleavages DNA containg thimine glycol adducts in deinococcus radiodurans. *Rad. Res.* 138, 282-285.

Murray R G E (1986). Family II. Deinococcaceae, p. 1035-1043. In P.H.A. Sneath, M.S.Mair, M.E. sharpe, and J.G. Holt (Ed.), Bergey's manual of systematic bacteriology, vol. 2. The Williams and Wilkins Co., Baltimore.

Murray RGE (1992). The family Deinococcaceae. In: The Prokaryotes, (Ed). A Ballows, HG Truper, M Dworkin, W Harder, KH Schleifer 4:3732–44. Springer-Verlag, New York USA.

Narumi I, Satoh K, Kikuchi M and Funayama T (2001). The LexA protein from *Deinococcus radiodurans* is not involved in RecA induction following □ irradiation. *J. Bacteriol.* 183 (23), 6951-6956.

Origanti D R, Pathak R, Dhaker A S, Arora R, Kumar R, Rajaram R and Gautam H K (2011). Isolation, Charecterization and bioactivity of deep sea bacteria with special reference to induction of antibacterial and antioxidant metabolites following gamma irradiation. *Canadian J Pure Appl. Sci.* 5(1), 1363-1370.

Punita S J, Reddy M A, and Das H K (1989). Multiple chromosome of *Azotobacter vinelandii*. *J. Bacteriol.* 171, 3133-3138.

Rakesh OD, Pathak R, Dhaker A S, Arora R, Kumar R, Rajaram R and Gautam H K (2011): Isolation, Charecterization and bioactivity of deep sea bacteria with special reference to induction of antibacterial and antioxidant metabolites following gamma irradiation. *Canadian J. Pure Appl. Sci.* 5(1), 1363-1370.

Rainey F A, Nobre M F, Schumann P, Stackebrandt E, Da-Costa M S (1997). Phylogenetic diversity of the Deinococci as determined by 16S ribosomal DNA sequence comparison. *Int. J. Syst. Bacteriol.* 47(2), 510-514.

Rouault T A and Klausner R D (1996). Iron sulfur clusters as biosensors of oxidants and iron. *Trends Biochem. Sci.* 21, 174-177.

Sadeff H L, Shimel B and Ellis S (1979). Charecterization of *Azotobacter vinelandii* deoxyribonucleic acid and folded chromosomes. *J. Bacteriol.* 136, 871-877.

Sauer U, Canonaco F, Heri S, Perrenoud A, Fischer E (2004). The soluble and membrane-bound transhydrogenases UdhA and PntAB have divergent functions in NADPH metabolism of *Escherichia Coli*. *J. Biol. Chem.* 279(8), 6613-6619.

Shibutani S, Takeshita M and Grollman A P (1991). Insertion of specific bases during DNA synthesis past the oxidation-damaged base 8-oxodG. Natu*re.* 349, 431-434.

Sun B, Latham KA, Dodson M L and Lioyd R S (1995). Studies on the catalytic mechanism of five DNA glycosylase: probing for enzyme-DNA imino intermediates. *J. Biol. Chem.* 270, 19501-19508.

Tchou J and Grollman AP (1993). Repair of DNA containg the oxidatively-damages base 8-Oxoguanine. Mutant Res. 299, 277-287.

Udupa K S, O'Cain P A, Mattimore V and Battista J R (1994). Novel ionizing radiation-sensitive mutants of *Deinococcus radiodurans*. *J. Bacteriol.* 176, 7439-7446.

Weisburg W G, Giovannoni S J and Woese C R (1989). The *Deinococcus-Thermus* phylum and the effect of rRNA composition on phylogenetic tree construction. *Syst. Appl. Microbiol.* 2, 128-134.

Wierowski J V and Bruce A K (1980). Modification of radiation resistance by manganese in *Micrococcus radiadurans. Radiat. Res.* 83, 384-389.

Wood M L Dizgaroglu M, Gajewski E and Essigmann J M (1990). Mechanistic studies of ionizing radiation and oxidative mutagenesis: effects of single 8-hydroxyguanine (7-hydro-8-oxoguanine) residue inserted at a unique abasic site in a viral genome. *Biochem.* 26, 7024-7032.

Wood M L, Dizdaroglu M, Gajewski E and Essigmann J M (1990). Mechanistic studies of ionizing radiation and oxidative mutagenesis: genetic effects of single 8-hydroxyguanine (7-hydro-8-oxyguanine) residue inserted at a unique site in a viral genome. *Biochem.* 29, 7024-7032.

In: Molecular Biology of Bacteria
Editors: R. Kumar Gaur and H. K. Gautam
ISBN: 978-1-62618-189-2
© 2013 Nova Science Publishers, Inc.

Chapter 10

Nitrogen Fixation Networking in *Pseudomonas Putida*

***Sakshi Issar*[1], *Hemant Gautam*[2], *D. K. Choudhary*[2] *and R. K.Gaur*[1,*]**
[1]Department of Science, faculty of Arts, Science and Commerce,
Mody Institute of Technology and Science-Lakshmangarh, Sikar, Rajasthan, India
[2]Institute of Genomics and Integrative Biology, Mall Road,
University Campus, Delhi, India

Abstract

Biological nitrogen fixation is brought about by free-living soil microorganisms and by symbiotic associations of microorganisms with higher plants. The nodules in leguminous plant become dwell for bacteria where they obtain energy from the host plant and take free nitrogen from the soil air and process it into combined nitrogen. In return, the plant receives the fixed N from nodules and produces food and forage protein. The bacterium's enzyme system supplies a constant source of reduced nitrogen to the host plant and the plant furnishes nutrients and energy for the activities of the bacterium. In the present study authors elaborately characterized the nitrogen network using the software CYTOSCAP that fixes nitrogen in genus *Pseudomonas*.

Keywords: Nitrogen fixation, *Pseudomonas putida*, metabolic networking, CYTOSCAPE

Introduction

The nutritional and catabolic versatility of *Pseudomonas* spp. undoubtedly contributes to their ubiquity in nature. Because of this advantage, especially in their aerobic competition for carbon (Tiedje, 1988), pseudomonads can be found in various habitats (e.g. soil, water, humans, animals, plants, and foods) where they can be saprophytic and may become

* Corresponding author's Email address: gaurrajarshi@hotmail.com.

beneficial or pathogenic to their hosts (Palleroni, 1984; Stanier *et al.*, 1966; Hugh and Gilardi, 1974; Stolp and Gadkari, 1981). They are one of the dominant groups of bacteria in marine (Quigley and Colwell, 1968) and soil (Paul and lark, 1989) systems including the rhizosphere where they can fix N_2 (Lifshitz et al., 1987; Krotzkyand Werner, 1987; Asanuma et al., 1979; Barraquio et al., 1983; Haahtela et al., 1983; Haahtela et al., 1983; Kleeberger et al., 1983; Klingmiiller, 1991; Prabha et al., 1978; Rahman et al., 1986; Thomas-Bauzon et al., 1982), produce plant growth-promoting substances (Lifshitz et al., 1987; Burr and Ceaser, 1984; Lalande, 1989) or excrete siderophores that exert antagonistic activity against soil-borne pathogens (De Weger et al., 1986; Schippers et al., 1987).

According to Palleroni (Palleroni, 1984), pseudomonads lack N_2-fixing ability. The ability of the pseudomonads to fix atmospheric N_2 has not been appreciated (Palleroni, 1984; Hill and Postgate, 1969; Meganathan, 1979) usually because of the following problems (Chan et al.,1986): (a) mistaken or uncertain identification of the diazotroph; (b) lack of rigorous tests for diazotrophy in pseudomonads; and (c) multigeneric nature of the genus *Pseudomonas* as discussed above. Early attempts to find an N_2-fixing pseudomonad were unsuccessful, due to the failure to recognise the O_2 sensitivity or microaerobic character of its N_2-fixing system. The existence of putative diazotrophic pseudomonads is now substantiated. Among these, *Pseudomonas* species strains 4B, DC and *P. pseudoflava* have been demonstrated to possess *nil* structural genes while *Pseudomonas* sp. 4B, *P. saccharophila* and *P. stutzeri* CMT.9.A have been shown to incorporate $^{15}N_2$. The genetically constructed N_2-fixing strains of *P. fluorescens* and *P. putida* indicate that these aerobic species have mechanisms to accommodate and express *nil* genes from *Klebsiella pneumoniae* although their maintenance may be a problem in the absence of selection pressure and protection from O_2 repression or inhibition (Mergeay and Gerits, 1978; Postgate and Kent, 1987).

Pathogenic Pseudomonads

Some pseudomonads can adopt a pathogenic lifestyle. The hosts for these pathogens include both plant and animals (including humans). Phytopathogens can cause disease to the most major groups of higher plants. The majority of them are foliar pathogens, although there are exceptions. The symptoms include necrotic lesions, twig dieback, hyperplasias, tissue maceration and vascular infections (Höfte and Vos, 2006; Schroth et al.,2006). The two most relevant representatives of plant pathogens are *Ralstonia solancearum* (formerly *Pseudomonas solacearum*), that cause wilts of many cultivated plants (potato, tomato, tobacco, peanut) and *Pseudomonas syryingae*, that attacks foliage and cause necrosis and necrotic lesions on leaves (Hayward, 1991; Höfte and Vos, 2006; Salanoubat *et.al.*,2002). The latter species has numerous pathovars, of which the hosts are usually limited to closely related species. *Burkholderia cepacia* (formerly *Pseudomonas cepacia*) and *Pseudomonas aeruginosa* are examples of strains that are able to infect animals. Both can be dangerous to humans, especially for those whose immune system functionality is impaired and are also common causes of nosocomial infections (Govan et al., 2007; Govan et al., 1996; Yahr and Parsek, 2006).

Role of *Pseudomonas* in N_2-Fixation

Pseudomonas-like organisms form one of the most complex groups of Gram-negative bacteria (Palleroni, 1984) apparently due to incomplete generic definition (De Vos*et. al.*, 1989). Bacteria which are aerobic, polarly flagellated, Gram-negative, non-spore-forming, rod-shaped, and which metabolize sugars oxidatively, but do not readily fit elsewhere, are often broadly classified as *Pseudomonas* spp. and they are usually differentiated by their phenotypic characteristics (Stanier *et. al.*, 1966) in conjunction withnumerical data analysis (Sneath and Sokal, 1973).

It now seems that several strains unambiguously classified as true *Pseudomonas* spp. can be added to the list of nitrogen fixers, on the basis of physiological properties, nitrogenase assays, phylogenetic studies and detection of nifH DNA by hybridization or PCR amplification (Chan *et. al.*, 1994; Vermeiren *et. al.*, 1999). Two nitrogen-fixing isolates belong to the species *Pseudomonas stutzeri. P. stutzeri* CMT.9.A was isolated from the roots of Sorghum (Krotzky and Werner 1987), whereas strain A15 originated from rice paddies in China (You and Zhou, 1989; You *et. al.*, 1991). Strain A15 was initially identified as an Alcaligenes faecalis, and was later reassigned to *P. stutzeri* (Vermeiren *et. al.*, 1999).

Putative N_2-fixing pseudomonad strains have been identified as belonging to eight acknowledged *Pseudomonas* species (*P.diminuta, P. fluorescens, P. paucimobilis* (proposed new name Sphingornonas paucimobilis (Yabuuchi *et. al.*, 1990), *P. pseudoflata, P. putida, P. saccharophila, P. stutzeri and P. tesicularis*) listed in Bergey's Manual of Systematic Bacteriology and fall into the rRNA groupsI, III and IV (Palleroni, 1984). Of these, *P. fluorescens* (Florenzano and Materassi, 1966), *P. putida* (Lifshitz *et. al.*, 1987; Lifshitz*et. al.*, 1986), and *P. stutzeri* strains JM300 (Barraquio *et. al.*, 1988; Carlson *et. al.*, 1983) and CMT.9.A (Barraquio *et. al.*, 1988; Krotzkyand Werner, 1987) belong to approved species in rRNA homology group I, which is kept in the genuine *Pseudomonas* genus classified under the Proteobacteria gamma subdivision (rRNA super family II) according to molecular phylogeny (De Vos*et. al.*, 1989; De Ley, 1992; Palleroni, 1992).

N_2 fixation by a strain resembling P.flava was reported by Berlier and Lespinat (1978) but no taxonomic evidence was given. In fact, the presence of N_2fixers within the genus *Pseudomonas* was not accepted until 1984 (Palleroni, 1984). However, N2 fixation by various strains related to acknowledged species of *Pseudomonas* was subsequently reported: P. saccharophila ATCC 15946 (Barraquio and Padreb, 1986); P. stutzeri CMT.9.A (Krotzky and Werner, 1987); *P. diazotrophicus* (Watanabe *et. al.*, 1987).

Metabolic Significance of *Pseudomonas*

Pseudomonas is a bacterial genus characterized by its metabolic versatility, genetic plasticity and ability to inhabit a variety of environments (Spiers *et. al.*, 2000). Together with the genera *Azotobacter, Azomonas, Azorhizophilus, Cellvibrio, Mesophilobacter, Rhizobacter, Rugamonas*, and *Serpens* it forms the family of Pseudomonadaceae that belongs to group of gamma-proteobacteria (Garrity, 2008). These bacteria are characterized by a chemoorganotrophic metabolism, their ability to grow on a broad range of organic substrates, the inability to perform photosynthesis and the absence of fermentation (Moore *et. al.*, 2006).

The genus *Pseudomonas* was established in 1895 by W. Migula (Migula, 1894) and since then it underwent many rearrangements, as numerous species originally assigned to, or at least suspected to be, *Pseudomonas*, were reclassified to different taxa in the course of further investigations (Matsumoto *et. al.*, 1978; Palleroni and Moore. 2004). To cope with this wealth of species, the term "pseudomonads" has been widely used in the scientific literature to describe bacteria that through their physiology and lifestyle resemble the genus *Pseudomonas*, although, according to current taxonomic knowledge, they cannot assigned to *Pseudomonas* or even to Pseudomonadaceae (Moore *et. al.*, 2006). This term will be used throughout this work, whenever the described properties can be attributed to the whole group of pseudomonads and not only to the *Pseudomonas* genus. Currently, this genus comprises 181 species (http://www.dsmz.de/microorganisms/bacterial_nomenclature.php) and it is further divided into a fluorescent and a non-fluorescent subgroup. Members of the former group are able to produce iron siderophores such as pyoverdin or pyocyanin under iron starvation, while the members of the latter are not (Palleroni and Doudorof, 1972). The members of *Pseudomonas* are gram-negative rods possessing one or more polar flagella that enable motility. The majority of the *Pseudomonas* species are obligate aerobes, while some species have been found to be facultative anaerobes. The latter can use nitrate respiration and/or arginine or pyruvate fermentation (Davies *et. al.*, 1989; Eschbach *et. al.*,2004; Haas *et. al.*, 1984; Zumft, 1997). *Pseudomonas* is a ubiquitous genus, colonizing a range of habitats ranging from water (including oligothropic mineral water) and soil to plant and animal tissues (Bergan, 1981; Moore *et. al.*, 2006; Schroth *et. al.*, 2006). Its members are also found in food sources where they act as spoiling agents (Barrett *et. al.*, 1986). As different as the environments are also the lifestyles, from saprophytes through parasites to pathogens. Some of the species are able to form biofilms, which complicates their removal if their presence is undesirable such as in infection settings (Costerton *et. al.*, 1995). In addition to an extensive metabolic versatility, pseudomonads also exhibit a number of interesting metabolic features, such as the ability to convert glucose extracellularly to gluconate and 2- oxogluconate, which renders the carbon inaccessible for many competitors. Due to the lack of phosphofructokinase, they use the Entner-Doudoroff pathway to metabolize glucose (Perron *et. al.*, 2004; Stanier *et. al.*, 1966). Many pseudomonads are also able to produce polyhydroxyalkanoates when growth is limited by a resource other than carbon, e.g. nitrogen (Moore *et. al.*, 2006; Rahman and Clarke, 1980). Pseudomonads have been attracting a considerable amount of interest in the scientific community due both to the biotechnological relevance of many of the strains, and the pathogenicity of some other.

Systems Biology – A Global Approach to Biology

By the most common definition Systems Biology is a field of study that deals with the interactions between the components of biological systems (Kitano, 2002 a and b). Some scientists see it as a research paradigm opposed to the reductionist one (Sauer et al.,2007), which assumes that a system is nothing but a sum of its parts, and it has been known in philosophy and science from ancient times. This approach proved to be successful in many disciplines since then, but the assumption that the interactions between the parts can be neglected does not hold in many systems (Westerhoff and Palsson, 2004). System Biology assumes that a living organism is not a simple sum of parts, but that the interactions between various elements of a biological system define many of its properties. From the interactions, new properties may emerge, that are not apparent when individual elements are treated separately. The new properties ('emergent properties') are in the particular focus of the realm of Systems Biology (Aderem, 2005; Boogerd et al., 2005; Ideker et al., 2001; Kitano, 2002 a and b).

Metabolic network reconstruction and simulation allows for an in depth insight into comprehending the molecular mechanisms of a particular organism, especially correlating the genome with molecular physiology (Francke et al., 2005). A reconstruction breaks down metabolic pathways into their respective reactions and enzymes, and analyzes them within the perspective of the entire network. Examples of various metabolic pathways include glycolysis, Krebs cycle, pentose phosphate pathway. In simplified terms, a reconstruction involves collecting all of the relevant metabolic information of an organism and then compiling it in a way that makes sense for various types of analyses to be performed. The correlation between the genome and metabolism is made by searching gene databases, such as KEGG (Dräger et al., 2009), GeneDB (Francke et al., 2005), for particular genes by inputting enzyme or protein names. For example, a search can be conducted based on the protein name or the EC number (a number that represents the catalytic function of the enzyme of interest) in order to find the associated gene (Francke et al., 2005).

In conclusion, metabolic network reconstruction and simulation can be effectively used to understand how an organism or parasite functions inside of the host cell. For example, if the parasite serves to compromise the immune system by lysing macrophages, then the goal of metabolic reconstruction/simulation would be to determine the metabolites that are essential to the organism's proliferation inside of macrophages. If the proliferation cycle is inhibited, then the parasite would not continue to evade the host's immune system. A reconstruction model serves as a first step to deciphering the complicated mechanisms surrounding disease. The next step would be to use the predictions and postulates generated from a reconstruction model and apply it to drug delivery and drug-engineering techniques. Currently, many tropical diseases affecting third world nations are very inadequately characterized, and thus poorly understood. Therefore, a metabolic reconstruction and simulation of the parasites that cause the tropical diseases would aid in developing new and innovative cures and treatments.

Construction of N_2-Metabolism Interaction Network of *P.putida*

The field of systems biology in living organisms is emerging as a consequence of publicly-available genomic, transcriptomics, proteomics, and metabolomics datasets. These data give us the hope of understanding the molecular function of the organism, and being able to predict the consequences to the entire system of a perturbation in the environment, or a change in expression of a single gene. In order to understand the significance of this data, the functional relationships between the genes, proteins, and metabolites must be put into context.

Biochemical pathways are thought of as functional units of metabolic networks. As such, the definitions and characterizations of metabolic pathways allow for a detailed analysis of robustness, physiological capabilities, and other systemic features of these complex reaction networks. These emergent properties necessitate the development of clear and mathematically precise definitions for a metabolic pathway. Mathematical and systemic definitions of metabolic pathways have been proposed (Mavrovouniotis and Stephanopoulos, 1990; Liao et al., 1996; Karp et al., 1999; Schuster et al., 1999; Ouzounis and Karp, 2000; Schilling et al., 2000).

The database used in this work is Kyoto Encyclopedia of Genes and Genomes (KEGG (Kanehisa and Goto, 2000), which contains databases of metabolic networks, which focus on prokaryotic organisms. KEGG computerizes current knowledge of molecular and cellular biology in terms of the interacting genes or molecules and links the pathways with the gene catalogs being produced by the genome sequencing projects. Cytoscape visualizes existing molecular interaction networks and gene expression profiles and other state data using Java (Shannon et al., 2003). Cytoscape has facilities for constructing networks and displaying annotations from fixed files. The central organizing metaphor of Cytoscape is a network graph, with molecular species represented as nodes and intermolecular interactions represented as links, that is, edges, between nodes. Cytoscape's Core software component provides basic functionality for integrating arbitrary data on the graph, a visual representation of the graph and integrated data, selection and filtering tools, and an interface to external methods implemented as plug-ins.

Bacteria in the genus *Pseudomonas* are ubiquitous inhabitants of the plant environment, where they colonize three distinct environments: foliar surfaces, plant tissues (particularly the intercellular spaces between plant cells), and the rhizosphere (Beattie and Lindow, 1999; Lugtenberg et al., 2001; Lindow and Brandl, 2003; Preston, 2004; Rico et al., 2008). Different species of *Pseudomonas* vary in their ability to thrive in different plant environments, with bacteria belonging to the species *Pseudomonas syringae* being well adapted for growth on leaf surfaces and in plant tissues, and bacteria belonging to the species *P. fluorescens, P. aeruginosa,* and *P. putida* showing a greater ability to thrive in the rhizosphere (Lugtenberg et al., 2001; Lindow and Brandl, 2003;Walker et al., 2004). Experimental analyses of nutrient assimilation by pseudomonads have shown that plant pathogenic *P. syringae* strains assimilate a restricted range of nutrient sources compared with other pseudomonads, particularly in terms of amino acids (Rico et al., 2008). This suggests that *P. syringae* is nutritionally specialized for growth using metabolites that are abundant in plant tissues and on foliar surfaces. The diversity of pseudomonads and the availability of genome sequence data for multiple plant-associated *P. syringae, P. putida,* and *P. fluorescens* strains, along with genome data for clinical isolates of *P. aeruginosa* and for the insect pathogen *P. entomophila* provide an excellent opportunity to use comparative genomic approaches to develop insight into the evolution of metabolic networks. However, although a few studies have developed metabolic models for individual strains (Nogales et al., 2008; Oberhardt et al., 2008), or compared the distribution of enzymes or protein domains across different species (Studholme et al., 2005; Perumal et al., 2008), most genome-based analyses of pseudomonads have focused on traits such as pathogenicity, gene expression, or degradation of xenobiotics (Collmer et al., 2002; Nelson et al., 2002; Feil et al., 2005; Joardar et al., 2005 a; b; Vencato et al.,2006; Loper and Gross, 2007; Lindeberg et al., 2008; Lindeberg et al., 2009; Silby et al., 2009). One approach that can be used to compare the metabolic properties of genome-sequenced organisms is to use data from publicly available data repositories such as Kyoto Encyclopedia of Genes and Genomes (KEGG) (Kanehisa et al., 2008) and MetaCyc and BioCyc (Caspi et al., 2008)to compare metabolic networks across genomes (Ma and Zeng, 2003; Light et al., 2005; Forst et al., 2006; Wang et al.,2006). However, due to the complex nature of metabolic networks, and the limited availability of experimentally validated phenotypic information, most comparative analyses have been restricted to analyzing and summarizing network properties, such as degree distribution, clustering coefficient, and average path length, or comparing small well-characterized regions within metabolic networks.

Pseudomonas putida is one of the best studied species of themetabolically versatile andubiquitous genus of the Pseudomonads(Timmis, 2002; dos Santos et al.,2004; Moore et al.,2006). As a species, it exhibits a wide biotechnological potential, with numerous strains (some of which solvent-tolerant (Mosqueda et al.,1999; de Bont, 1998)able to efficiently produce a range of bulk and fine chemicals. These features, along with their renowned stress resistance, amenability for genetic manipulation and suitability as a host for heterologous expression, make *Pseudomonas putida* particularly attractive for biocatalysis. To date, strains of *P. putida* have been employed to produce phenol, cinnamic acid, cis-cis-muconate, p-hydroxybenzoate, p-cuomarate, and myxochromide (Wierckx et al., 2005; Nijkamp et al., 2005; Choi et al., 1997; Ramos-Gonzalez et al., 2003; Verhoef *et. al.*, 2007; Nijkamp et al., 2007; Stephan *et. al.*, 2006). Furthermore, enzymes from *P. putida* have been employed in a variety of other biocatalytic processes, including the resolution of D/L-phenylglycinamideinto D-phenylglycinamide and L-phenylglycine, productionof non-proteinogenic L-amino acids, and biochemicaloxidation of methylated heteroaromaticcompounds for formationof heteroaromatic monocarboxylic acids (Schmid *et. al.*, 2001). However, most *Pseudomonas*-based applications are still in infancy largely due to a lack of knowledge of the genotype-phenotype relationships in these bacteria under conditions relevant for industrial and environmental endeavors.

Biological nitrogen fixation, a key step in global nitrogen cycling, is the ATP-dependent reduction of dinitrogen to ammonia by the nitrogenase enzyme complex. The ability to fix nitrogen is widely distributed among *Bacteria* and *Archaea*. In addition to symbiotic nitrogen-fixing microorganisms, which induce differentiated structures on the host plant (root nodules of legumes and actinorhizal plants), there exist a variety of free-living nitrogen-fixing bacteria capable of association with the root system of graminaceous plants, such as *Klebsiella pneumoniae*, *Azotobacter vinelandii*, and *Azospirillum brasilense* (Elmerich and Newton, 2007). New species, such as *Azoarcus* and *Gluconacetobacter*, can also develop endophytic associations and survive within plant tissues without causing disease symptoms (Elmerich and Newton, 2007; Krauseet al., 2006). Sakshi (2011) constructed the gene interaction network of *P. putida* manually with the help of CYTOSCAPE software (Figure 1) and compare with the Rhizobium sp. It is universally accepted that *Rhizobium* are the main bacteria who exclusively participate in N_2-fixation with legumes without harming the plants, also beneficial in form of biofertilizers and also show PGP activities. But exceptions are always there that it is not obligatory that it could be found everywhere. For example- due to harsh climate there is found very squat population of *Rhizobia* in Deserts. Although many researches shows the presence of this bacteria even in deserts but only in vegetative parts like- Aravalli regions in Rajasthan (Rathore et al., 2009). Now a day with advances in biotechnology, it's become easy to culture the *Rhizobium* in *in vitro* conditions which is usefulin sustainable agriculture. The main point come in focus was that on the place of nifH gene known as nitrogenase reductase, which take active part in N2-fixation in *Rhizobium*, was not found in *Pseudomonas putida*. Nitrogenase was the gene product in *P.putida*.

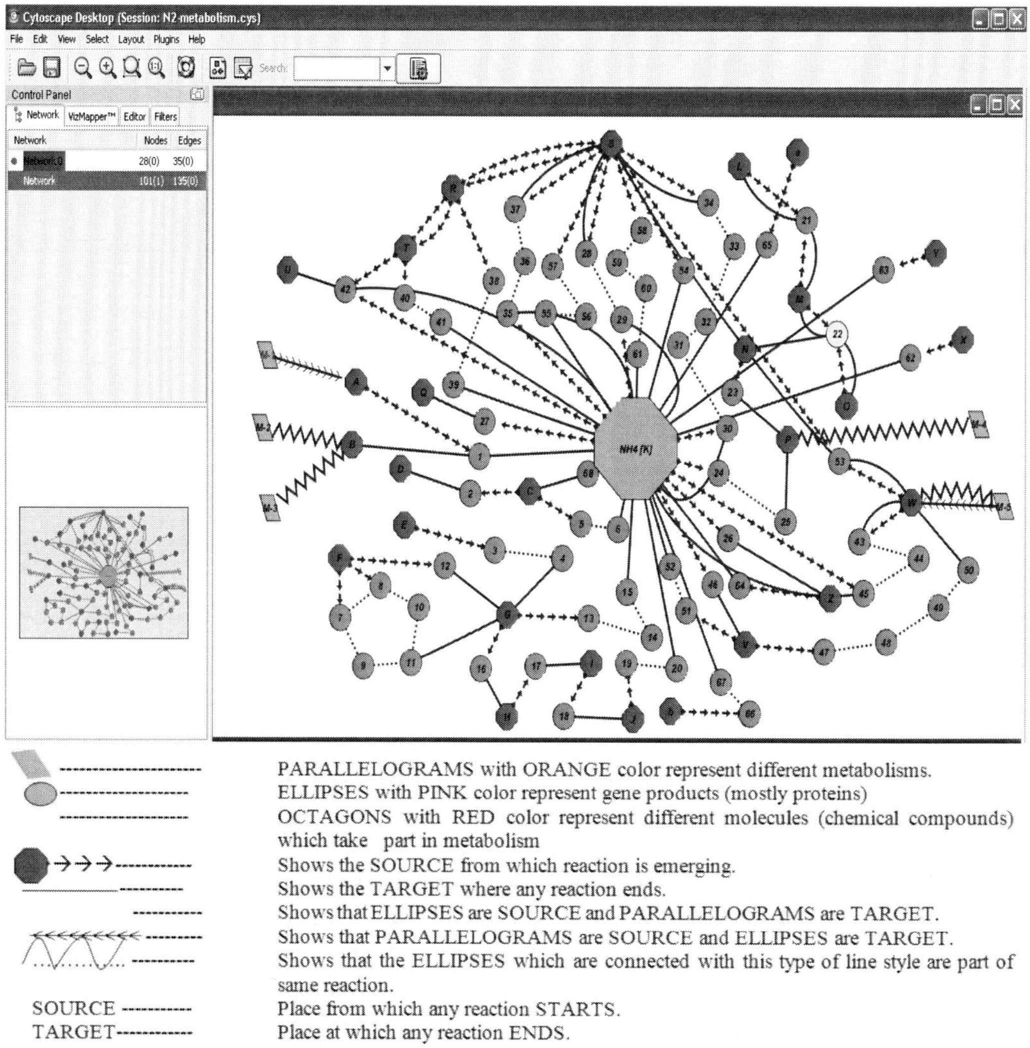

Figure 8. A gene interaction network of N_2-Metabolism developed by Cytoscape Software (*P.putida*).

Acknowledgments

The authors are thankful to Department of Science and Technology, Govt. of India, Rajasthan for providing the financial support under student for the present study.

References

Aderem, A. (2005). Systems biology: Its practice and challenges. *Cell,* 121:511-513.

Asanuma, S., Tanaka, H. and Yatazawa, M. (1979). Rhizoplane microorganisms of rice seedlings as examined by scanning electron microscopy. *Soil Sci. Plant Nutr.,* 25:539-551.

Barraquio, W.L., Dumont, A. and Knowles, R. (1988). Enumeration of free-living aerobic N_2-fixing H_2-oxidizing bacteria by using a heterotrophic semisolid medium and most-probable-number technique. *Appl. Environ. Microbiol.*, 54:1313-1317.

Barraquio, W.L., Ladha, J.K. and Watanabe, I. (1983). Isolation and identification of N_2-fixing *Pseudomonas* associated with wetland rice. *Can. J. Microbiol.*, 29:867- 873.

Barraquiwo, L. and Padreb, C. (1986).Nitrogen fixation by *Pseudomonas* saccharophila Doudoroff ATCC15946. *Journal of General Microbiology,* 132:237-241.

Barrett, E.L., Solanes, R.E., Tang, J.S. and Palleroni, N.J. (1986).*Pseudomonas fluorescens* biovar-V -Its Resolution into Distinct Component Groups and the Relationship of These Groups to Other *Pseudomonas-Fluorescens* biovars, to *Pseudomonas putida*, and to Psychrotrophic Pseudomonads Associated with Food Spoilage. *Journal of General Microbiology,* 132:2709-2721.

Beattie, G.A. and Lindow, S.E. (1999). Bacterial colonization of leaves: a spectrum of strategies. *Phytopathology*, 89:353–359.

Bergan, T. (1981). Human-and animal-pathogenic members of the genus *Pseudomonas*, *In* Starr, M.P., Stolp, H., Trüper, H.G., Balows, A. and Schlegel H.G. (eds.). The Prokaryotes, vol. 1. Springer-Verlag, Berlin. pp. 666-700.

Berlierj, M. and Lespinatp, A. (1978). Relationship between nitrogenase and hydrogenase activity in a rhizosphere diazotroph. *FEMS Microbiology Letters,* 3:187-190.

Boogerd, F.C., Bruggeman, F.J., Richardson, R.C., Stephan, A. and Westerhoff, H.V. (2005). Emergence and its place in nature: a case study of biochemical networks. *Synthese*, 145:131-164.

Burr, T.J. and Ceaser, A. (1984). Beneficial plant bacteria. *Crit. Rev. Plant Sci.*, 2:1-20.

Carlson, C.A., Pierson, L.S., Rosen, J.J. and Ingraham, J.U. (1983).*Pseudomonas stutzeri* and related species undergo natural transformation. *J. Bacteriol.*, 153:93-99.

Caspi, R., Foerster, H., Fulcher, C.A. et al., (13 co-authors) (2008). The MetaCyc Database of metabolic pathways and enzymes and the BioCyc collection of Pathway/Genome Databases. *Nucleic Acids Res.,* 36:D623–D631.

Chan, Y. K., Barraquio, W. L. and Knowles, R. (1994). N_2-fixing pseudomonads and related soil bacteria. *FEMS Microbiol. Rev.*, 13: 95–118.

Chan, Y.-K., Wheatcroft, R. and Watson, R.J. (1986).Physiological and genetic characterization of a diazotrophic pseudomonad. *J. Gen. Microbiol.*, 132:2277- 2285.

Choi, W.J., Lee, E.Y., Cho, M.H. and Choi C.Y. (1997). Enhanced production of cis, cis muconate in a cell-recycle bioreactor. *J. Ferment. Bioeng.,* 84: 70–76.

Collmer, A., Lindeberg, M., Petnicki-Ocwieja, T., Schneider, D.J. and Alfano, J.R. (2002). Genomic mining of type III secretion system effectors in *Pseudomonas* syringae yields new picks for all TTSS prospectors. *Trends Microbiol.,* 10:462–469.

Costerton, J.W., Lewandowski, Z., Caldwell, D.E., Korber, D.R. and Lappinscott, H.M. (1995). Microbial Biofilms. *Annual Review of Microbiology,* 49:711-745.

Davies, K.J.P., Lloyd, D. and Boddy, L. (1989). The Effect of Oxygen on Denitrification in *Paracoccus denitrificans* and *Pseudomonas aeruginosa*. *Journal of General Microbiology,* 135:2445-2451.

De Bont, J.A.M. (1998). Solvent-tolerant bacteria in biocatalysis. *Trends Biotechnol.,* 16:493–499.

De Ley, J. (1992). The proteobacteria: ribosomal RNA cistron similarities and bacterial taxonomy. In: The Prokaryotes, Vol. 2, Ballows, A., Triiper, H.G., Dworkin, M., Ttarder, W. and Schleifer, K.H. (eds.). SpringerVerlag, New York, NY, pp. 2111-2140.

De Vos, P., van Landschoot, A., Segers, P., *et. al.*, (1989). Genotypic relationships and taxonomic localization of unclassified *Pseudomonas* and *Pseudomonas-like* strains by deoxyribonucleic acid: ribosomal ribonucleic acid hybridizations. *Int. J. Syst. Bacteriol.*, 39:35-49.

De Weger, L.A., van Boxtel, R., van der Burg, B., Gruyters, R.A, Schippers, B. and Lugtenberg, B. (1986). Siderophores and outer membrane proteins of antagonistic, plant-growth-stimulating, root-colonizing *Pseudomonas* spp. *J. Bacteriol.*, 165:585-594.

dos Santos V.A.P.M., Heim S., Moore E.R.B., Stratz M. and Timmis K.N. (2004). Insights into the genomic basis of niche specificity of *Pseudomonas putida* KT2440. *Environ Microbiol.,* 6:1264–1286.

Dräger, A., Kronfeld, M., Ziller, M. J., Supper, J., Planatscher, H., Magnus, J. B., Oldiges, M., Kohlbacher, O. and Zell, A. (2009). Modeling metabolic networks in *C. glutamicum*: a comparison of rate laws in combination with various parameter optimization strategies. *BMC Systems Biology.*, 3(5).

Elmerich, C. and Newton, W.E. (2007). Associative and Endophytic Nitrogen fixing Bacteria and Cyanobacterial Associations—Introduction. *Springer,* Dordrecht, The Netherlands.

Eschbach, M., Schreiber, K., Trunk, K., Buer, J., Jahn, D. and Schobert, M. (2004). Long-term anaerobic survival of the opportunistic pathogen *Pseudomonas* aeruginosa via pyruvate fermentation. *Journal of Bacteriology,* 186:4596-4604.

Feil, H, Feil, WS, Chain, P., et al., (20 co-authors) (2005). Comparison of the complete genome sequences of *Pseudomonas* syringae pv. syringae B728a and pv. tomato DC3000. *Proc. Natl. Acad. Sci. USA,* 102:11064–11069.

Florenzano, G. and Materassi, R. (1966). Morphological, physiological and biochemical study of some new nitrogen- fixing strains of *Pseudomonas (P.* fluorescens var. indologenes, nov. var. and P. fluorescens var. vallis umbrosae, nov. var.) isolated from forest soils. *Ann. Microbiol. Enz.,* 16:187-200.

Forst, C., Flamm, C., Hofacker, I. and Stadler, P. (2006). Algebraic comparison of metabolic networks, phylogenetic inference, and metabolic innovation. *BMC Bioinformatics,* 7:67.

Francke, C., Siezen, R. J. and Teusink, B. (2005). Reconstructing the metabolic network of a bacterium from its genome. *Trends in Microbiology*, 13(11):550-558.

Garrity, G.M. (2008). Bergey's Manual of Systematic Bacteriology. *Springer.*

Govan, J.R.W., Hughes, J.E. and Vandamme, P. (1996).*Bur*kholderia cepacia: Medical, taxonomic and ecological issues. *Journal of Medical Microbiology*, 45:395-407.

Govan, J.R., Brown, A.R. and Jones, A.M. (2007). Evolving epidemiology of *Pseudomonas* aeruginosa and the Burkholderia cepacia complex in cystic fibrosis lung infection. *Future Microbiology*, 2:153-164.

Haahtela, K., Helander, I., Nurmiaho-Lassila, E.L. and Sundman, V. (1983). Morphological and physiological characteristics and lipopolysaccharide composition of N_2-fixing (C_2H_2-reducing) root-associated *Pseudomonas* sp. *Can. J. Microbiol.*, 29:847-880.

Haahtela, K., Kari, K. and Sundman, V. (1983). Nitrogenase activity (acetylene reduction) of root-associated, cold-climate *Azospirillum,* Enterobacter, Klebsiella, and *Pseudomonas* species during growth on various carbon sources and at various partial pressures of oxygen. *Appl. Environ. Microbiol.*, 45:563-570.

Haas, D., Matsumoto, H., Moretti, P., Stalon, V. and Mercenier, A. (1984). Arginine Degradation in *Pseudomonas* aeruginosa Mutants Blocked in 2 Arginine Catabolic Pathways. *Molecular and General Genetics.,* 193:437-444.

Hayward, A.C. (1991). Biology and Epidemiology of Bacterial Wilt Caused by *Pseudomonas solanacearum*. *Annual Review of Phytopathology*, 29:65-87.

Hill, S. and Postgate, J.R. (1969). Failure of putative nitrogen-fixing bacteria to fix nitrogen. *J. Gen. Microbiol.,* 58:277-285.

Höfte, M. and Vos, P. (2006). Plant pathogenic *Pseudomonas* species.In:Gnanamanickam S. S. (eds.). Plant-Associated Bacteria. *Springer,* pp. 507-533.

Hugh, R. and Gilardi, G.L. (1974).*Pseudomonas.* In: Manual of Clinical Microbiology Lennette, E.H., Spaulding, E.H. and Truant, J.P. (eds.).*American Society for Microbiology, Washington, D.C.*pp. 250-269,

Ideker, T., Galitski, T. and Hood, L. (2001). A new approach to decoding life: Systems biology. *Annual Review of Genomics and Human Genetics,* 2:343-372.

Joardar, V, Lindeberg, M, Jackson, R.W., et al., (32 co-authors) (2005). Whole-genome sequence analysis of *Pseudomonas* syringae pv. phaseolicola 1448, A reveals divergence among pathovars in genes involved in virulence and transposition. *J. Bacteriol.,* 187:6488–6498.

Joardar, V., Lindeberg, M., Schneider, D.J., Collmer, A. and Buell, C.R. (2005). Lineage-specific regions in *Pseudomonas syringae* pv. tomato DC3000. *Mol. Plant. Pathol.,* 6:53–64.

Kanehisa, M., Araki, M., Goto, S., et al., (11 co-authors) (2008). KEGG for linking genomes to life and the environment. *Nucleic Acids Res.,* 36:D480–D484.

Kanehisa, M. and Goto, S. (2000). Kegg: Kyoto Encyclopedia of Genes and Genomes, *Nucleic Acids Research,* 28(1):27-30.

Karp, P.D., Krummenacker, M., Paley, S. and Wagg, J. (1999). Integrated pathway-genome databases and their role in drug discovery. *Trends Biotech.,* 17:275–281.

Kitano, H. (2002). Computational systems biology. *Nature*, 420:206-210.

Kitano, H. (2002). Systems biology: A brief overview. *Science*, 295:1662-1664.

Krause, A., et al., (2006). Complete genome of the mutualistic, N_2-fixing grass endophyte *Azoarcus* sp. *strain* BH72. *Nat. biotechnol.,* 24:1385–1391.

Lalande, R., Bissonnette, N., Coutl6e, D. and Antoun, H. (1989). Identification of rhizobacteria from maize and determination of their plant-growth promoting potential. *Plant Soil*, 115:7-11.

Liao, J.C., Hou, S.Y. and Chao, Y.P. (1996). Pathway analysis, engineering and physiological considerations for redirecting central metabolism. *Biotech. Bioengr,* 52:129–140.

Lifshitz, R., Kloepper, J.W., Kozlowski, M., Simonson, C., Carlson, J., Tipping, E.M. and Zaleska, I. (1987). Growth promotion of canola (rapeseed) seedlings by a strain of *Pseudomonas putida* under gnotobiotic conditions. *Can. J. Microbiol.*, 33:390-395.

Lifshitz, R., Kloepper, J.W., Scher, F.M., Tipping, E.M. and Lalibert, M. (1986). Nitrogen-fixing pseudomonads isolated from roots of plants grown in the Canadian High Arctic. *Appl. Environ. Microbiol.*, 51:251-255.

Light, S., Kraulis, P. and Elofsson, A. (2005). Preferential attachment in the evolution of metabolic networks. *BMC Genomics,* 6:159.

Lindeberg, M., Biehl, B., Glasner, J., Perna, N., Collmer, A. and Collmer, C. (2009). Gene Ontology annotation highlights shared and divergent pathogenic strategies of type III effector proteins deployed by the plant pathogen *Pseudomonas* syringae pv. tomato DC3000 and animal pathogenic Escherichia coli strains. *BMC Microbiol.,* 9:S4.

Lindeberg, M., Myers, C.R., Collmer, A. and Schneider, D.J. (2008). Roadmap to new virulence determinants in *Pseudomonas* syringae: insights from comparative genomics and genome organization. *Mol. Plant Microbe Interact.,* 21:685–700.

Lindow, S.E. and Brandl, M.T. (2003). Microbiology of the phyllosphere. *Appl. Environ Microbiol.,* 69:1875–1883.

Loper, J. and Gross, H. (2007). Genomic analysis of antifungal metabolite production by *Pseudomonas* fluorescens Pf-5. *Eur. J. Plant Pathol.,* 119:265–278.

Lugtenberg, B.J.J, Dekkers, L. and Bloemberg, G.V. (2001). Molecular determinants of rhizosphere colonization by *Pseudomonas*. *Annual Review of Phytopathology,* 39:461–490.

Ma, H., and Zeng, A.P. (2003). Reconstruction of metabolic networks from genome data and analysis of their global structure for various organisms. *Bioinformatics*, 19:270–277.

Matsumoto, H., Ohta, S., Kobayashi, R. and Terawaki, Y. (1978). Chromosomal Location of Genes Participating in Degradation of Purines in *Pseudomonas aeruginosa*. *Molecular and General Genetics,* 167:165-176.

Mavrovouniotis, M.L. and Stephanopoulos, G. (1990). Computer-aided synthesis of biochemical pathways. *Biotech. Bioengr.* 36:1119–1132.

Meganathan, R. (1979). Why has the 'nitrogenase plasmid' snubbed the pseudomonads. *J. Theor. Biol.,* 80:301-304.

Mergeay, M. and Gerits, J. (1978). F'-plasmid transfer from *Escherichia coli* to *Pseudomonas fluorescens*. *J. Bacteriol.*, 135:18-28.

Migula, W. (1894). Over a new system of the bacteria. *Work Bakteriol Inst. Karlsruhe*, 1:235-238.

Moore, E.R.B., Tindall, B.J., Martins dos Santos, V.A.P, Pieper, D.H., Ramos, J.L. and Palleroni, N.J. (2006). Nonmedical: *Pseudomonas*. In: The Prokaryotes: A Handbook on the Biology of Bacteria. Dworkin, M., Falkow, S., Rosenberg, E., Schleifer, K., Stackebrandt, E. (eds.) New York: *Springer*, pp. 646–703.

Mosqueda, G., Ramos-Gonzalez, M.I. and Ramos J.L. (1999). Toluene metabolism by the solvent-tolerant *Pseudomonas* putida DOT-T1 strain, and its role in solvent impermeabilization. *Gene*, 232:69–76.

Nelson, K.E., Weinel, C., Paulsen, I.T., *et. al.*, (43 co-authors) (2002). Complete genome sequence and comparative analysis of the metabolically versatile *Pseudomonas putida* KT2440. *Environ Microbiol.,* 4:799–808.

Nijkamp, K., van Luijk, N., de Bont, J.A.M. and Wery, J. (2005). The solvent-tolerant *Pseudomonas* putida S12 as host for the production of cinnamic acid from glucose. *Appl. Microbiol. Biotechnol.,* 69:170–177.

Nijkamp, K., Westerhof, R.G.M., Ballerstedt, H., de Bont., J.A.M. and Wery, J. (2007). Optimization of the solvent-tolerant *Pseudomonas* putida S12 as host for the production of p-coumarate from glucose. *Appl. Microbiol. Biotechnol.,* 74:617–624.

Nogales, J., Palsson, B.O. and Thiele, I. (2008). A genome-scale metabolic reconstruction of *Pseudomonas* putida KT2440: iJN746 as a cell factory. *BMC Syst. Biol.*, 2:79.

Oberhardt, M.A., Puchalka, J., Fryer, K.E,. Martins dos Santos, V.A.P. and Papin, J.A. (2008). Genome-scale metabolic network analysis of the opportunistic pathogen *Pseudomonas* aeruginosa PAO1. *J. Bacteriol.,* 190:2790–2803.

Ouzounis, C.A. and Karp, P.D. (2000). Global properties of the metabolic map of *Escherichia coli. Genome Res.,* 10:568–576.

Palleroni, N. J. and M. Doudorof. (1972). Some Properties and Taxonomic Subdivisions of Genus *Pseudomonas. Annual Review of Phytopathology,* 10:73-100.

Palleroni, N. J. and Moore, E. R. B. (2004). Taxonomy of Pseudomonads: experimental approaches, In:Ramos, J. L. (eds.).*Pseudomonas*, vol. 1. Kluwer Academic/Plenum Publishers, New York, pp. 3-44.

Palleroni, N.J. (1984). Genus I. *Pseudomonas* (Migula 1894). In: Krieg, N.R., Holt, J.G. (eds.). Bergey's manual of systematic bacteriology. Baltimore, MD: Williams and Wilkins, pp. 141–199.

Palleroni, N.J. (1992).Present situation of the taxonomy of aerobic pseudomonads. In: *Pseudomonas:* Molecular Biology and Biotechnology. In: Galli, E., Silver, S. and Witholt, B., (eds.).American Society for Microbiology, Washington, D.C, pp. 105-115,

Paul, E.A. and Clark, F.E. (1989). Soil Microbiology and Biochemistry. Academic Press, San Diego, CA, pp. 49-73.

Perron, K., Caille, O., Rossier, C., van Delden, C., Dumas, J. L. and Kohler, T. (2004). CzcR-CzcS, a two-component system involved in heavy metal and carbapenem resistance in *Pseudomonas aeruginosa. Journal of Biological Chemistry*, 279:8761-8768.

Perumal, D., Lim, C.S., Chow, V.T.K., Sakharkar, K.R. and Sakharkar, M.K. (2008). A combined computational-experimental analyses of selected metabolic enzymes in *Pseudomonas* species. *Int. J. Biol. Sci.,* 4:309–317.

Postgate, J.R. and Kent, H.M. (1987). Qualitative cvidence for expression of *Klebsiella pneumoniae nif* in *Pseudomonas putida. J. Gen. Microbiol.*, 133:2563-2566.

Prabha, S., Nair, S.K., Dadarwal, K.R. and Tauro, P. (1978).*Pseudomonas* azotogensis - an associated symbiont of Bajra (Pennisetum americanum). *Plant Soil,* 49:657-659.

Preston, G.M. (2004). Plant perceptions of plant growth-promoting *Pseudomonas. Philos. Trans. R Soc. B,* 359:907–918;

Quigley, M.M. and Colwell, R.R. (1968). Properties of bacteria isolated from deep-sea sediments. *J. Bacteriol*, 95:211-220.

Rahman, M. and Clarke, P. H. (1980). Genes and Enzymes of Lysine Catabolism in *Pseudomonas aeruginosa. Journal of General Microbiology,* 116:357-369.

Rahman, M., Weinhard, P., Bally, R., Hubert, J.L. and Balandreau, J. (1986). New nitrogen-fixing *Pseudomonas* species in the dominant rhizosphere bacterial population of rice. In: Fourth Int. Symp. *Microb. Ecol, Ljubljana*, pp. 115.

Ramos-Gonzalez, M.I., Ben-Bassat, A., Campos, M.J. and Ramos, J.L. (2003). Genetic engineering of a highly solvent-tolerant *Pseudomonas* putida strain for biotransformation of toluene to p-hydroxybenzoate. *Appl. Environ. Microbiol.,* 69:5120–5127.

Rathore, M.S., Shekhawat, N.S. and Gehlot, H.S. (2009).Need of Assessing *Rhizobia* for Their Plant Growth Promoting Activities Associated with Native Wild Legumes Inhabiting Aravalli Ranges of Rajasthan, India.*Botany Research International*, 2(2):115-122.

Rico, A. and Preston, G.M. (2008).*Pseudomonas syringae* pv. Tomato DC3000 uses constitutive and apoplast-induced nutrient assimilation pathways to catabolize nutrients that are abundant in the tomato apoplast. *Mol. Plant Microbe Interact.,* 21:269–282.

Salanoubat, M., Genin, S., Artiguenave, F., Gouzy, J., Mangenot, S., Arlat, M., Billault, A., Brottier, P., Camus, J.C., Cattolico, L., Chandler, M., Choisne, N., Claudel-Renard, C.,

Cunnac, S., Demange, N., Gaspin, C., Lavie, M., Moisan, A., Robert, C., Saurin, W., Schiex, T., Siguier, P., Thebault, P., Whalen, M., Wincker, P., Levy, M., Weissenbach, J. and Boucher, C.A. (2002). Genome sequence of the plant pathogen Ralstonia solanacearum. *Nature*, 415:497-502.

Sauer, U., Heinemann, M. and Zamboni, N. (2007). GENETICS: Getting Closer to the Whole Picture. *Science*, 316:550.

Schilling, C.H., Letscher, D. and Palsson, B.O. (2000). Theory for the systemic definition of metabolic pathways and their use in interpreting metabolic function from a pathway-oriented perspective. *J. Theor. Biol.,* 203:229–248.

Schippers, B., Bakker, A.W. and Bakker, P.A.H.M. (1987). Interactions of deleterious and beneficial rhizosphere microorganisms and the effect of cropping practices. *Ann. Rev. Phytopathol.*, 25:339-358.

Schmid, A., Dordick, J.S., Hauer, B., Kiener, A., Wubbolts, M., et al., (2001). Industrial biocatalysis today and tomorrow. *Nature*, 409:258–268.

Schroth, M.N., Hildebrand, D.C. and Panopoulos, N. (2006). Phytopathogenic Pseudomonads and Related Plant-Associated Pseudomonads.In:Dworkin, M., Falkow,S., Rosenberg, E., Schleifer, K.and Stackebrandt E. (eds.). The Prokaryotes. *Springer,* pp. 714-740.

Schuster, S., Dandekar, T. and Fell, D.A. (1999). Detection of elementary flux modes in biochemical networks: A promising tool for pathway analysis and metabolic engineering. *Trends Biotechnol,* 17:53–60.

Shannon, P., Markiel, A., Ozier, O., Baliga, N., Wang, J., Ramage, D., Amin, N., Schwikowski, B. and Ideker, T. (2003).Cytoscape: A Software Environment for Integrated Models of Biomolecular Interaction Networks.*Genome Research,* 13(11):2498-504.

Silby, M.W., Cerdeno-Tarraga, A.M., Vernikos, G., et al., (31 co-authors) (2009). Genomic and functional analyses of diversity and plant interactions of *Pseudomonas fluorescens. Genome Biol.,* 10:R51.

Sneath, P.H.A. and Sokal, R.R. (1973). Numerical Taxonomy. W.H. Freeman, San Francisco, CA.

Spiers, A. J., Buckling, A. and Rainey, P. B. (2000). The causes of *Pseudomonas* diversity. *Soc. General. Microbiol.,* 144:2345-2350.

Stanier, R.Y., Palleron, N.J. and Doudorof, M. (1966). Aerobic Pseudomonads - a Taxonomic Study. *Journal of General Microbiology*, 43:159-271.

Stephan, S., Heinzle, E., Wenzel, S.C., Krug, D., Muller, R, *et. al.*, (2006). Metabolic physiology of *Pseudomonas* putida for heterologous production of myxochromide. *Process Biochem.,* 41:2146–2152.

Stolp, H. and Gadkari, D. (1981). Nonpathogenic members of the genus *Pseudornonas.* In: The Prokaryotes, Vol. 1, Starr, M.P., Stolp, H., Triiper, H.G., Balows, A. and Schlegel, H.G. (eds.). Springer-Verlag, Berlin, pp. 719-741.

Studholme, D., Downie, A. and Preston, G. (2005). Unique domain architectures in plant pathogenic Proteobacteria. *BMC Genomics,* 6:17.

Thomas-Bauzon, D., Weinhard, P., Villecourt, P. and Balandreau, J. (1982). The spermosphere model. I. Its use in growing, counting and isolating N_2-fixing bacterial from the rhizosphere of rice. *Can. J. Microbiol.*, 28:922-928.

Tiedje, J.M. (1988). Ecology of denitrification and dissimilatory reduction to ammonium. In: Biology of Anaerobic Microorganisms, Zehnder A.J.B. (eds.).*Wiley*, New York, *NY*, pp. 179-244.

Timmis, K.N. (2002).*Pseudomonas* putida: a cosmopolitan opportunist par excellence. *Environ. Microbiol.*, 4:779–781.

Vencato, M., Tian, F., Alfano, J.R.,et al., (15 co-authors) (2006). Bioinformatics-enabled inventory of the *Hrp* regulon and type III secretion system effector proteins of *Pseudomonas* syringae pv. phaseolicola 1448a. *Mol. Plant Microbe Interact.,* 19:1193–1206.

Verhoef, S., Ruijssenaars, H.J., de Bont, J.A.M. and Wery, J. (2007). Bioproduction of phydroxybenzoate from renewable feedstock by solvent-tolerant *Pseudomonas* putida S12. *J. Biotechnol.,* 132:49–56.

Vermeiren, H., Willems, A., Schoofs, G., de Mot, R., Keijers, V., Hai, W. and Vanderleyden, J. (1999). The rice inoculant strain A15 is a nitrogen-fixing *Pseudomonas stutzeri* strain. *Syst. Appl. Microbiol.,* 22:215–224.

Walker, T.S., Bais, H.P., Deziel, E., Schweizer, H.P., Rahme, L.G., Fall, R. and Vivanco, J.M. (2004).*Pseudomonas* aeruginosa-plant root interactions. Pathogenicity, biofilm formation, and root exudation. *Plant Physiol.,* 134:320–331.

Wang, Z., Zhu, X.G., Chen, Y., Li, Y., Hou, J., Li Y. and Liu, L. (2006). Exploring photosynthesis evolution by comparative analysis of metabolic networks between chloroplasts and photosynthetic bacteria. *BMC Genomics,* 7:100.

Watanabei, S.R., Ladha, J.K., Katayamafujimuray,and Huraishi, H. (1987).A new nitrogen-king species of pseudomonad : Pseudomnas diazotrophicus sp. nov. isolated from the root of wetland rice. *Canadian Journal of Microbiology,* 33:670-678.

Westerhoff, H. V. and Palsson, B. O. (2004). The evolution of molecular biology into systems biology. *Nature Biotechnology*, 22:1249-1252.

Wierckx, N.J.P, Ballerstedt, H., de Bont, J.A.M. and Wery, J. (2005). Engineering of solvent-tolerant *Pseudomonas putida* S12 for bioproduction of phenol from glucose. *Appl. Environ. Microbiol.*, 71:8221–8227.

Yabuuchi, E., Yanok, I., Oyaizu, H., Hashimoto, Y., Ezaki, T. and Yamamoto, H. (1990). Proposals of Sphingomonas paucimobilis gen. nov. and comb. nov., Sphingomonas yanoikuyae sp. nov., Sphingomonas capsulate comb. nov., and two genospecies of the genus *Sphingomonas. Microbiol. Immunol.*, 34:99-119.

Yahr, T.L. and Parsek, M.R. (2006).*Pseudomonas aeruginosa*.InDworkin,M., Falkow,S., Rosenberg,E., Schleifer, K.and Stackebrandt, E. (eds.). The Prokaryotes. *Springer*, pp. 704-713.

You, C. and Zhou, F. (1989). Non-nodular endorhizospheric nitrogen fixation in wetland rice. *Can. J. Microbiol.*, 35:403–408.

You, C. B., Song, H. X., Wang, J. P., Lin, M. and Hai, W. L. (1991). Association of Alcaligenes faecalis with wetland rice. *Plant Soil*, 137:81–85.

Zumft, W. G. (1997). Cell biology and molecular basis of denitrification. *Microbiology and Molecular Biology Reviews*, 61:533-616.

In: Molecular Biology of Bacteria
Editors: R. Kumar Gaur and H. K. Gautam
ISBN: 978-1-62618-189-2
© 2013 Nova Science Publishers, Inc.

Chapter 11

Quorum Sensing in *Pseudomonas Aeruginosa*: Cell to Cell Signaling at Molecular Level

Sakshi Issar and R. K. Gaur [*]
Department of Science, Faculty of Arts, Science and Commerce,
Mody Institute of Technology and Science, Lakshmangarh, Sikar, Rajasthan, India

Abstract

Quorum sensing is a generic regulatory mechanism which allows bacteria to begin a joined and synchronized response in a population density-dependent manner to achieve different tasks. The bacteria synthesize and secrete different chemical signals which can either be produced at different stages of growth or can be cell-density related (autoinducers). These signals allow bacteria to scrutinize their environment and modify gene expression to obtain a competitive advantage. Among pseudomonads, *P. aeruginosa* is used as a model organism to study the cell-to cell signaling at pathogenic level. The quorum sensing network of *P. aeruginosa* is made up of 2 interlinked N-acyl homoserine lactone (AHL) dependent circuits and numerous regulators, which make it highly complex network as compare to quorum sensing networks of other bacteria. In this article the quorum sensing system of *P. aeruginosa* has been summarized which can be proved a powerful tool to solve many biological problems.

Keywords: Quorum sensing, *Pseudomonas aeruginosa,* N-acyl homoserine lactone

Introduction

Quorum Sensing in prokaryotic biology refers to the ability of a bacterium to sense information from other cells in the population when they reach a critical concentration (i.e. a

[*] Corresponding author's Email address: gaurrajarshi@hotmail.com.

Quorum). The phenomenon of Quorum sensing or cell- to- cell communication relies on the principle that when a single bacterium releases autoinducers (AIs) in the environment, their concentration is too low to be detected. However, when sufficient bacteria are present, AI concentrations reach a threshold level that allows the bacteria to sense a critical cell mass and, in response, to activate or repress target genes (Vasil, 2003).

Cell-to-cell communication has been proven to play an important role in the life of various bacterial species since the discovery of population density-dependent regulation of bioluminescence in marine bacteria. Past few year's work provided evidence that bacteria form well organized communities and exchange information with the other members of the community in order to coordinate their activities (Duan et al., 2003; Greenberg, 2003; Ben Jacob et al ., 2004; Eber and Tümmler, 2004).

This allows a group of bacteria to launch a unified, coordinated response to environmental stimuli and to accomplish tasks which would be difficult, if not impossible, to achieve for a single bacterial cell. Furthermore, evidence has accumulated that some bacterial signal molecules are used not only as population density sensors in one species but also for communication between bacteria of different species or genera occupying the same ecological niche and even to interact with their eukaryotic hosts. Beside bioluminescence and genetic competence, quorum sensing is also involved in the regulation of a wide variety of different physiological processes including antibiotic biosynthesis, swarming, swimming and twitching motility, plasmid conjugal transfer, biofilm development or the production of bacterial virulence factors in plant, animal or human pathogens (Miller and Bassler, 2001; Whitehead et al ., 2001; Camara et al ., 2002; Fuqua and Greenberg, 2002; Lazdunski et al ., 2004; Pappas et al ., 2004).

Quorum-sensing systems were originally discovered in the marine bacterium *Vibrio fischeri* as being involved in the control of light production (Nealson, 2003), and have been identified in a wide range of bacterial genera, including *Pseudomonas*, *Escherichia*, and *Streptococcus*. Although bacteria can detect a signal at any concentration, autoinducer-mediated communication often occurs in a concentration-dependent manner. This population effect is utilized by bacteria to regulate phenotypes in a manner that enables them to adapt and survive continually changing environments by coupling individual cell responses to population-wide alterations. On the other hand, other bacterial signals can also be utilized in cell-to-cell communication without being involved in population-wide regulation of traits and phenotypes.

The origin of these signals differs—some signals (e.g., indole) may be produced by bacteria as part of their normal metabolism, whereas others (e.g., hydroxyindole) may be produced by bacteria modifying these signals and utilizing them for regulating different phenotypes (Luo et al., 2003; Luo and Morrison, 2003).

One of the most extensively studied AHL-dependent cell-to-cell communication systems is the one of the opportunistic human pathogen *Pseudomonas aeruginosa*. This bacterium, well known for its resistance against a variety of antibiotics, causes serious infections in immune-compromised patients or individuals with cystic fibrosis and is one leading source of nosocomial infections (Tümmler et al., 1991). Work of the past few years showed that the quorum sensing circuitry that is operating in *P. aeruginosa* is essential for the expression of virulence factors as well as for biofilm formation (Hentzer et al., 2004) and it has been speculated that quorum sensing- dependent regulation of pathogenic traits is a highly effective strategy to establish an infection (de Kievit and Iglewski, 2000; Greenberg, 2003).

This review is primarily focused on cell-to-cell communication in *P. aeruginosa* mediated by quorum sensing. We first discussed the quorum sensing system of first originally discovered gram negative marine bacterium *Vibrio fischeri* and then quorum sensing system of *P. aeruginosa* and its regulation is discussed in detail along with quorum sensing in host-pathogen interaction and other pseudomonads.

Quorum Sensing System in Gram Negetive Bacteria

AHL-mediated Quorum sensing systems are studied in a large number of Gram Negative bacterial species belonging to the α, β and γ subclasses of proteobacteria, including bacteria in the genera *Agerobacterium, Aeromonas, Burkholderia, Chromobacterium, Citrobacter, Enterobacter, Ervinia, Hafnia, Nitrosomonas, Obesumbacterium, Pantoea, Pseudomonas, Rahnella, Ralstonia, Rhodobacter, Rhizobium, Serretia, Vibrio, Xenorhabdus* and *Yersinia* (Table-1). Among the Gram negative bacteria, the most well- studied Quorum sensing system is the LuxR- LuxI homologous system and the cognate signal molecules: *N*- acyl- homoserine lactones (AHLs) (Kempner et al., 1968). When in high enough concentration, these molecules can bind to and activate a transcriptional activator, or R protein, which in turn induces expression of target genes (Figure 1).

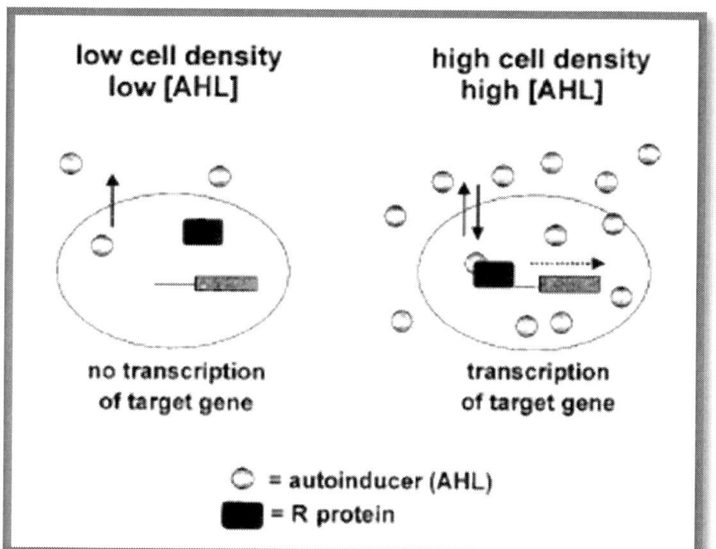

Figure 1. Quorum sensing in gram-negative organisms involves two regulatory components: the transcriptional activator protein (R protein) and the AI molecule produced by the autoinducer synthase. Accumulation of AI occurs in a cell-density-dependent manner until a threshold level is reached. At this time the AI binds to and activates the R protein, which in turn induces gene expression. The R protein consists of two domains: the N terminus of the protein that interacts with AI and the C terminus that is involved in DNA binding. Typically, gram-negative AI molecules are *N*-acyl-HSLs; however, other types of signal molecules do exist.

Quorum Sensing System in P. Aeruginosa

With regard to bacteria that utilize quorum sensing as part of their pathogenic lifestyle, *P. aeruginosa* is perhaps the best understood in terms of the virulence factors regulated and the role quorum sensing plays in pathogenicity. Classified as an opportunistic pathogen, *P. aeruginosa* primarily infects individuals who are immune-compromised, such as patients with cancer or AIDS (Fergie et al., 1994; Mendelson et al., 1994) or those having breaches in normal barriers caused by burns, indwelling medical devices, or prolonged use of broad-spectrum antibiotics (Brewer et al., 1996; Dunn and Wunderink, 1995).

Table 1. Summary of quorum sensing in gram-negative bacteria

Organism	**Major signal molecule**	**Regulatory proteins**	**Phenotype**
Vibrio fischeri	3-Oxo-C6-HSL	LuxI/LuxR	Bioluminescence
Vibrio harveyi	3-Hydroxy-C4-HSL	LuxLM/LuxN	Bioluminescence
	?	Lux?/LuxPQ	
Vibrio anguillarum	?	VanI/VanR	3-Oxo-C10-HSL
Pseudomonas aeruginosa	3-Oxo-C_{12}-HSL	LasI/LasR	Multiple extracellular enzymes, RhlR, Xcp, biofilm formation
	C4-HSL	RhlI/RhlR	Multiple extracellular enzymes, rhamnolipid, RpoS, secondary metabolites
Pseudomonas aureofaciens	C6-HSL	PhzI/PhzR	Phenazine antibiotics
Agrobacterium tumefaciens	3-Oxo-C8-HSL	TraI/TraR	Ti plasmid conjugation
Erwinia carotovora subsp. carotovora	3-Oxo-C6-HSL	ExpI/ExpR CarI/CarR	Exoenzymes Carbapenem antibiotics
Erwinia chrysanthemi	3-Oxo-C6-HSL	ExpI/ExpR	Pectate lyases
	C6-HSL		
Erwinia stewartii	3-Oxo-C6-HSL	EsaI/EsaR	Exopolysaccharide, virulence factors
R. leguminosarum	C6-HSL	RhiI/RhiR	RhiABC rhizosphere-expressed genes, nodulation
	C8-HSL		
	3-Hydroxy-7-cis-C_{14}-HSL		
Rhizobium etli	?	RaiI/RaiR	Restriction of number of nitrogenfixing nodules
Chromobacterium violaceum	C6-HSL	CviI/CviR	Exoenzymes, antibiotics, cyanide, violacein
Burkholderia cepacia	C8-HSL	CepI/R	Protease, siderophores
Aeromonas hydrophila	C4-HSL	AhyI/AhyR	Exoprotease production
Aeromonas salmonicida	C4-HSL	AsaI/AsaR	Extracellular protease
Ralstonia solanacearum	C8-HSL	SolI/SolR	?
Serratia liquifaciens	C4-HSL	SwrI/SwrR	Extracellular protease, swarming
Rhodobacter sphaeroides	7-cis-C_{14}-HSL	CerI/CerR	Dispersal from bacterial aggregates
Organism	Major signal molecule	Regulatory proteins	Phenotype
Enterobacter agglomerans	3-Oxo-C_6-HSL	EagI/EagR	?
E. coli	?	?/SdiA	Cell division, attachment and effacing lesion formation
Yersinia enterocholitica	C6-HSL	YenI/YenR	?
Y. pseudotuberculosis	C8-HSL	YesI/YesR	?

The genetic basis for this growth-phase regulation was uncovered with the discovery that *P. aeruginosa* contains genes, called *lasR* and *lasI*, with significant homology to the *luxR* and *luxI* genes of *Vibrio fischeri* (Gambello and Iglewski, 1991; Passador et al., 1993).

In *P. aeruginosa*, the transcriptional activator LasR functions in conjunction with its cognate AHL, *N*-(3-oxododecanoyl)-L-homoserine lactone (3-oxo-C_{12}-HSL), synthesized by the LasI autoinducer synthase (Passador et al., 1993**;** Pearson et al., 1994). LasR–3-oxo-C_{12}-HSL regulates expression of a number of *P. aeruginosa* virulence genes including *lasB*, *lasA*, *aprA*, and *toxA* (Gambello and Iglewski, 1991; Gambello et al., 1993; Pearson et al., 1994; Toder et al., 1991) as well as *lasI* itself, creating an autoinduction feedback loop (Seed et al., 1995) (Figure 2). An additional gene, *rsaL*, is under the regulatory control of LasR–3-oxo-C_{12}-HSL, the product of which negatively regulates *P. aeruginosa* quorum sensing by inhibiting *lasI* expression (de Kievit et al., 1999).

The second signaling *rhl* quorum-sensing system consists of the transcriptional activator RhlR and the autoinducer synthase RhlI which directs the synthesis of *N*-butyryl- L-homoserine lactone (C4-HSL) (Pearson *et. al.*, 1995). The RhlR–C4-HSL complex regulates expression of *rhlAB*, required for rhamnolipid production, *lasB*, *aprA*, the stationary-phase sigma factor RpoS, and production of the secondary metabolites pyocyanin and cyanide (Brint and Ohman 1995; Latifi *et. al.*, 1996; Latifi *et. al.*, 1995, Ochsner *et. al.*, 1994, Pearson *et. al.*, 1995; Winson *et. al.*, 1995).

With the finding that *P. aeruginosa* has two separate quorum- sensing circuits came the question of whether the two were capable of interaction. In spite of the predicted structural similarities between LasR and RhlR and the similarities between the two AHLs, there is little interchangeability between the two systems. The R-proteins are not significantly activated by their non-cognate AHLs; LasR is not activated by C4-HSL and 3-oxo-C_{12}-HSL is capable of only low-level RhlR activation (Pearson *et. al.*, 1997). Thus it appears that the R proteins show high specificity with regard to the AHL required for their activation. Similarly, genes that are primarily activated by one system are only minimally activated by the other (Pearson *et. al.*, 1997), indicating that specific recognition sequences must be present in the operator regions of these target genes that dictate which quorum-sensing system is required for induction. Despite the high fidelity of these systems for their regulatory components and gene targets, a link between the two systems does exist. The *las* system positively regulates expression of both *rhlR* and *rhlI* (Latifi et al., 1996; Pesci et al., 1997) (Figure 2). Furthermore, 3-oxo-C_{12}-HSL is able to compete with C4-HSL for RhlR binding, indicating that 3-oxo-C_{12}-HSL is able to act as an antagonist of the *rhl* system (Pesci et al., 1997). Thus, it appears that in *P. aeruginosa*, quorum sensing is arranged in a hierarchical fashion with the *las* system being the dominant regulator.

One possible role for these minor AHL molecules is to activate additional LuxR-type proteins. In *P. aeruginosa*, two genes encoding proteins with significant homology to LasR and RhlR have been identified; however, at this time it is unclear whether the minor signal molecules present in *P. aeruginosa* culture supernatants can activate either of these R proteins.

A second possible role for non-cognate AHLs arises from the fact that these molecules can frequently activate a given R protein, albeit at lower induction levels than for the cognate AHL. In this manner, minor AHLs may function as competitive inhibitors of autoinduction. An example of this is seen in *P. aeruginosa* where the *las* signal molecule 3-oxo-C_{12}-HSL can efficiently compete with C4-HSL for RhlR binding (Pesci et al., 1997).

Recently, a third autoinducer molecule was identified in *P. aeruginosa* (Pesci et al., 1999). This molecule is structurally very different from the other two *P. aeruginosa* autoinducers in that it is a 2-heptyl-3-hydroxy-4-quinolone, designated PQS. Preliminary studies have revealed that PQS is involved in *lasB* expression and that although expression of PQS is under control of the *las* system, RhlR is required for PQS activity. The structural similarity between PQS and antimicrobial quinolones is quite interesting, although preliminary studies have not shown any antimicrobial activity associated with PQS (Pesci et al., 1999).

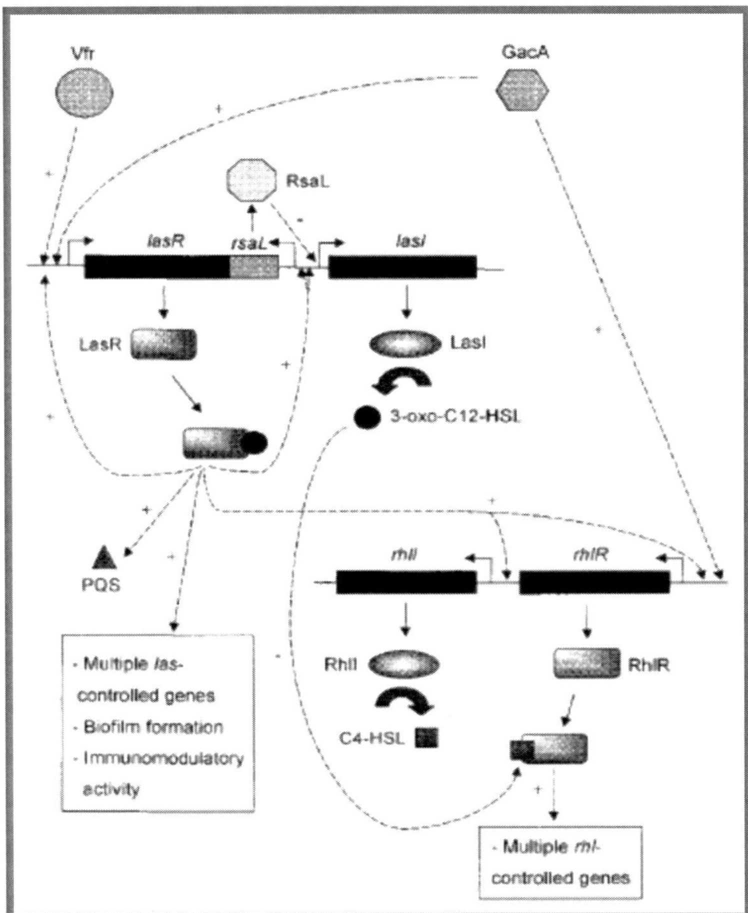

Figure 2. The quorum-sensing circuitry of *P. aeruginosa* is illustrated. Expression of the *lasR* gene is subject to at least two levels of control: the global regulators Vfr and GacA and the *las* quorum-sensing system, which regulates expression of both *lasR* and *lasI*. The latter creates an autoinduction feedback loop. Regulation of the *rhl* system is similar to *las* in that GacA affects expression of *rhlR* and the *rhlR* and *rhlI* genes are controlled to some degree by the *las* system. Interestingly, the *las* quorum-sensing system was shown to elicit an additional level of control over the *rhl* system; the *las* signal molecule, 3-oxo- C_{12}-HSL, can act post-translationally to block RhlR activation by C4-HSL. The *las* and *rhl* quorum-sensing systems regulate expression of numerous genes that contribute to the virulence of *P. aeruginosa*. In addition, the *las* signal molecule, 3-oxo-C_{12}-HSL, is required for biofilm differentiation and exhibits immune-modulatory activity.

The discovery of PQS reveals yet another layer in the increasingly complex system used by this organism to maintain tight control of its virulence factors. This tight regulation is a common theme in *P. aeruginosa* quorum sensing, evidenced by the fact that the *xcp* genes involved in type II secretion are under control of both the *las* and *rhl* quorum-sensing systems (Chapon-Herve *et. al.*, 1997). This pathway is utilized in secretion of quorum-sensing controlled enzymes, such as elastase and proteases, indicating that *P. aeruginosa* is extremely vigilant about regulating these factors at production and export levels.

P. aeruginosa is intrinsically resistant to numerous antimicrobial agents, including antibiotics, organic solvents, and detergents. Low outer membrane permeability together with the presence of multidrug efflux pumps that export a wide range of antimicrobial agents is thought to contribute to this intrinsic resistance. Three well-studied *P. aeruginosa* pumps have been described: MexAB-OprM, MexCD-OprJ, and MexEF-OprN encoded by the *mexAB-oprM*, *mexCD-oprJ*, and *mexEF-oprN* operons, respectively (Kohler *et. al.*, 1997; Poole et al., 1996; Poole *et. al.*, 1993).

During a study to investigate whether AIs freely diffuse in and out of *P. aeruginosa* cells, it was discovered that in addition to its slow diffusion, 3-oxo-C_{12}-HSL is actively pumped from cells by the MexAB-OprM pump (Pearson *et. al.*, 1999). In contrast, C4-HSL diffuses rapidly across the cell membranes and is not actively transported (Pearson *et. al.*, 1999).

Intriguingly, a *gacA* mutant and a *toxA* mutant also exhibited decreased virulence in the three models (Rahme *et. al.*, 1995; Rahme *et. al.*, 1997; Tan *et. al.*, 1999).

GacA is a global activator in *P. aeruginosa* that has previously been shown to regulate expression of *lasR* and *rhlR* and production of the *rhl* AHL, C4-HSL (Reimmann et al., 1997); *toxA* encodes exotoxin A, which is regulated by the *las* quorum-sensing system (Gambello *et. al.*, 1993). The aforementioned genes, which are all linked to quorum sensing, contribute to the trans-kingdom virulence of *P. aeruginosa*.

Regulation of Quorum Sensing in P. Aeruginosa

The quorum sensing cascade of *P. aeruginosa* is subject to regulation by a number of additional regulatory factors. One important factor is the two-component response regulator *gacA*, a highly conserved gene primarily responsible for the control of virulence factors and antibiotics production in fluorescent pseudomonads, but also shown to be involved in the regulation of C_4-HSL production in *P. aeruginosa* (Reimmann et al., 1997).

Likewise, the global transcriptional regulator AlgR2 (AlgQ) binds to the LasR and RhlR promoter regions thereby down regulating the entire quorum sensing cascade in *P. aeruginosa* (Ledgham et al., 2003a). RsaL repressor has an important role in the quorum sensing circuitry, as it competes with the LasR-3-oxo-C_{12}-HSL complex for the binding to the *lasI* promoter, thus negatively regulating *lasI* transcription (de Kievit et al., 1999; Fagerlind et al., 2004). Polyphosphate kinase, encoded by *ppk*, has a positive impact on 3-oxo-C_{12}-HSL and C_4-HSL synthesis and subsequently on the overall virulence of *P. aeruginosa* (Rashid et al., 2000).

A gene *mvaT*, when mutated, led to increased and premature expression of quorum sensing-regulated genes, indicating that MvaT is required for the appropriate timing of expression of the quorum sensing regulon (Diggle et al., 2002). Moreover, quorum sensing is regulated at post-transcriptional level by the RsmA/RsmZ system (Pessi et al., 2001; Heurlier et al., 2004) as well as by DksA (Jude et al., 2003).

The quorum sensing signalling network of virulence is linked with the cAMP-signalling network of virulence (Wolfgang et al., 2003) via the transcriptional regulator Vfr (Albus et al., 1997), a member of the cAMP receptor protein family. Vfr is important for the regulation of exotoxin A and protease, which is due in part to its influence on the expression of *lasR lasI*. In addition, Vfr positively regulates the expression of over a hundred genes, including those encoding the type III secretion system that delivers toxic effector proteins directly into host cells (Wolfgang et al., 2003).

In *P. aeruginosa* cAMP is primarily synthesized by CyaB, a class III membrane-associated adenylate cyclase, which together with Vfr have been shown to act as a master regulator of virulence gene expression in *P. aeruginosa* (Wolfgang et al., 2003; Smith et al., 2004). Bioinformatic analyses of the *P. aeruginosa* PAO1 genome revealed that, in addition to the two AHL signal receptors LasR and RhlR, the *P. aeruginosa* genome encodes two additional LuxR homologues, termed QscR and VqsR. The first LuxR homologue, QscR, whose synthesis is regulated by GacA, has been shown to modulate the timing of quorum sensing gene expression by repressing transcription of *lasI* (Chugani et al., 2001).

In the absence of AHLs, QscR forms multimers interacting with itself and heterodimers with LasR or RhlR, thus offering another possibility of fine tuning the quorum sensing circuitry in concert with other regulatory proteins (Ledgham et al., 2003b). Inactivation of the second identified LuxR homologue, VqsR, abrogates production of both signalling molecules 3-oxo-C_{12}-HSL and C4-HSL, decreased production of virulence factors and reduced the pathogenicity of *P. aeruginosa*. On the other hand, the presence of the *las* box located in the promoter upstream of *vqsR* places this gene under the direct control of the *las* quorum sensing system (Juhas et al., 2004).

The alternative sigma factor RpoN shown negatively affect quorum sensing in *P. aeruginosa*. The inactivation of *rpoN* was found to raise the levels of 3-oxo-C_{12}- HSL (Heurlier et al., 2003). However, depending on the environmental conditions, the expression of RhlI mRNA transcript, encoding C4-HSL can be either increased or decreased by RpoN (Heurlier et al., 2003; Thompson et al., 2003). Furthermore, the stationary sigma factor RpoS was demonstrated to be involved in the modulation of expression of a large number of quorum sensing genes as well as the type III secretion system and anti-host effector proteins (Whiteley et al., 2000; Hogardt et al., 2004; Schuster et al., 2004). Vice versa, RhlR was shown to regulate expression of this sigma factor (Latifi et al., 1996), thus suggesting mutual control between quorum sensing and RpoS. All the regulatory systems mentioned add additional levels of complexity to quorum sensing circuitry of *P. aeruginosa* (Figure-3).

According to genome-wide microarray-based transcriptome analyses, approximately 5% of all *P. aeruginosa* genes are quorum sensing-regulated. Quorum sensing in *P. aeruginosa* was shown to modulate expression of a broad spectrum of extracellular proteins, secondary metabolites, and regulatory proteins as well as many proteins belonging to the class of conserved hypothetical of unknown function (Hentzer et al., 2004; Schuster et al., 2003; Wagner et al., 2003; Juhas et al., 2004). A transcriptome analysis of the *vqsR* mutant provided clear evidence that VqsR is involved in the regulation of the expression of certain sets of virulence, quorum sensing and iron-regulated genes (Cornelis and Aendekerk, 2004; Juhas et al., 2004).

In a recent study the quorum sensing regulon of sessile *P. aeruginosa* cells was mapped by transcriptomics and in this study many genes were identified, which were previously shown to be iron-regulated (Hentzer et al., 2004).

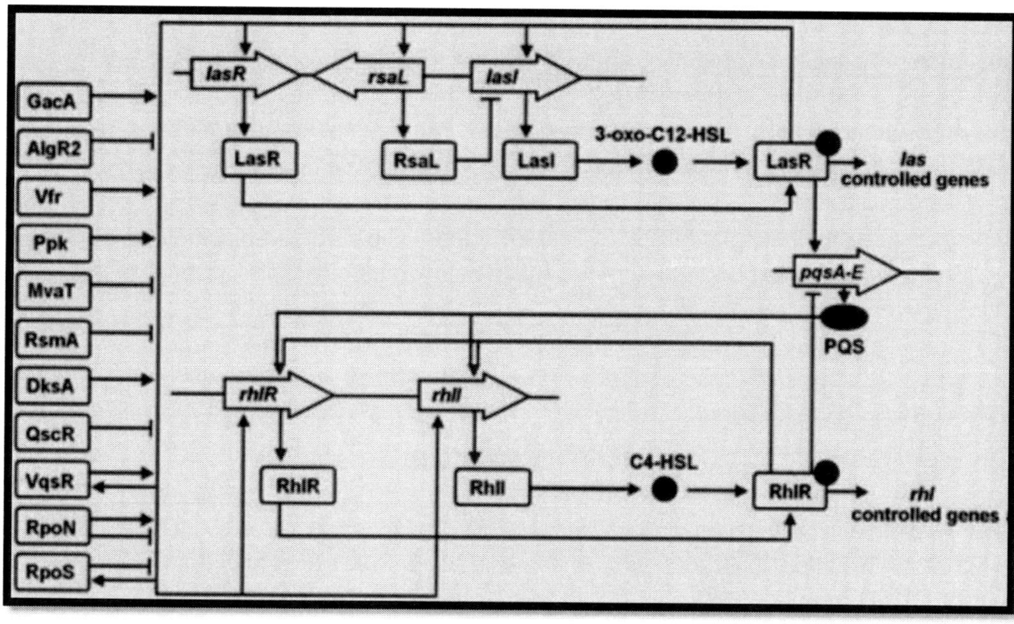

Figure 3. The quorum sensing network in *P. aeruginosa*. The two hierarchically arranged *las* and *rhl* quorum sensing circuits are subject to modulation by a number of additional regulators, which fine tune the quorum sensing response of the organism (activation= inhibition= ——┤).

Quorum Sensing in Host–Pathogen Interactions

Quorum sensing was found to modulate the expression of a wide variety of bacterial virulence genes (Passador et al., 1993). The multiple extracellular pathogenic traits regulated by quorum sensing in *P. aeruginosa* include elastases, alkaline protease, exoenzyme S, neuraminidase, haemolysin, lectins, pyocyanin, rhamnolipids, hydrogen cyanide or oxidative stress-responsive enzymes catalase and superoxide dismutase (Passador et al., 1993; Latifi et al., 1995; Pearson et al., 1997; Hasset et al., 1999; Pessi and Haas, 2000; Winzer et al., 2000) (Table-1). All these extracellular virulence factors are crucial for the competence of *P. aeruginosa* to establish and maintain the infection. Mutants defective in quorum sensing are typically compromised in their ability to establish a successful infection. The importance of quorum sensing for the pathogenicity of *P. aeruginosa* has been demonstrated in a number of animal models. Quorum sensing mutants showed reduced virulence in a number of non-mammalian infection models, including *Caenorhabditis elegans*, *Dictyostelium discoideum* and *Arabidopsis thaliana* (Rahme et al., 1995; Tan et al., 1999; Cosson et al., 2002) in the burned mouse model (Tang et al., 1996; Rumbaugh et al., 1999; Wu, Y.P. et al., 2000).

Conclusion

In the field of Microbiology, Cell-to-cell communication represents one of the most rapid evolving fields of research. Since the discovery of cell-to-cell communication in the marine bacterium *V. fischeri*, the quorum sensing mechanism has been proved a generic feature of many bacterial species. A complex quorum sensing signalling network was identified in *P. aeruginosa,* made up of two AHLs and one non-AHL-related signal and numerous transcriptional regulators (Figure 5) plays a key role in the regulation of a battery of extracellular virulence factors and which represents an attractive target for the design of novel anti-pseudomonal drugs. AHLs can be produced by genetically modified plants and further can be used to manipulate plant–bacterium associations. The ability to block or promote these QS systems provides a powerful tool to solve many problems and enhance productivity including the development of synthetic gene networks and modules, antimicrobial therapeutics, and biotechnology applications.

References

Albus, A. M., Runyen-Janecky, L. J., West, S. E. H. and Iglewski, B. H. (1997). Vfr controls quorum sensing in *Pseudomonas aeruginosa*. *J. Bacteriol,* 179, 3928–3935.

Atkinson, S., Throup, J. P., Stewart, G. S. and Williams, P. (1999). A hierarchical quorum-sensing system in *Yersinia pseudotuberculosis* is involved in the regulation of motility and clumping. *Mol. Microbiol,* 33, 1267–1277.

Bainton, N. J., Stead, P., Chhabra, S. R., Bycroft, B. W., Salmond, G. P. C., Steward, G. S. A. B. and Williams, P. (1992). *N*-(3-Oxohexanoyl)-L-homoserine lactone regulates carbapenem antibiotic production in *Erwinia carotovora*. *Biochem. J.,* 288, 997–1004.

Bassler, B. L., Wright, M. and Silverman, M. R. (1994). Multiple signalling systems controlling expression of luminescence in *Vibrio harveyi*: sequence and function of genes encoding a second sensory pathway. *Mol. Microbiol,* 13, 273–286.

Bassler, B. L., Wright, M., Showalter, R. E. and Silverman, M. R. (1993). Intercellular signalling in *Vibrio harveyi*: sequence and function of genes regulating expression of luminescence. *Mol. Microbiol,* 9, 773–786.

Beck von Bodman, S. and Farrand, S. K. (1995). Capsular polysaccharide biosynthesis and pathogenicity in *Erwinia stewartii* require induction by a *N*-acyl-homoserine lactone autoinducer. *J. Bacteriol,* 177, 5000–5008.

Ben Jacob, E., Becker. I., Shapira, Y. and Levine, H. (2004). Bacterial linguistic communication and social intelligence. *Trends Microbiol,* 12, 366–372.

Brewer, S. C., Wunderink, R. G., Jones, C. B. and Leeper, K. V. J. (1996). Ventilator-associated pneumonia due to *Pseudomonas aeruginosa*. *Chest*, 109, 1019–1029.

Brint, J. M. and Ohman, D. E. (1995). Synthesis of multiple exoproducts in *Pseudomonas aeruginosa* is under the control of RhlR-RhlI, another set of regulators in strain PAO1 with homology to the autoinducer-responsive LuxR-LuxI family. *J. Bacteriol,* 177, 7155–7163.

Camara, M., Williams, P. and Hardman, A. (2002). Controlling infection by tuning and turning down the volume of bacterial small-talk. *Lancet Infect Dis,* 2, 667–676.

Cao, J. and Meighen, E. A. (1989). Purification and structural identification of an autoinducer for the luminescence system of *Vibrio harveyi*. *J. Biol. Chem.*, 264, 21670–21676.

Chapon-Herve, V., Akrim, M., Latifi, A., Williams, P., Lazdunski, A. and Bally, M. (1997). Regulation of the *xcp* secretion pathway by multiple quorum sensing modulons in *Pseudomonas aeruginosa*. *Mol. Microbiol.* 24, 1169–1178.

Chhabra, S. R., Stead, P., Bainton, N. J., Salmond, G. P. C., Stewart, G. S. A. B., Williams, P. and Bycroft, B. W. (1993). Autoregulation of carbapenem biosynthesis in *Erwinia carotovora* ATCC 39048 by analogues of *N*-3-(oxohexanoyl)-L-homoserine lactone. *J. Antibiot,* 46, 441–454.

Chugani, S.A., Whiteley, M., Lee, K.M., D'Argenio, D., Manoil, C. and Greenberg, E.P. (2001). QscR, a modulator of quorum-sensing signal synthesis and virulence in *Pseudomonas aeruginosa*. *Proc. Natl. Acad. Sci. USA,* 98, 2752–2757.

Cornelis, P. and Aendekerk, S. (2004). A new regulator linking quorum sensing and iron uptake in *Pseudomonas aeruginosa*. *Microbiology,* 150, 752–756.

Cosson, P., Zulianello, L., Join-Lambert, O., Faurisson, F., Gebbie, L., Benghezal, M., et al. (2002). *Pseudomonas aeruginosa* virulence analysed in a *Dictyostelium discoideum* host system. *J. Bacterio.,* 184, 3027–3033.

Cubo, M. T., Economou, A., Murphy, G., Johnston, A. W. B. and Downie, J. A. (1992). Molecular characterization and regulation of the rhizosphere-expressed genes *rhiABCR* that can influence nodulation by *Rhizobium leguminosarum* biovar *viciae*. *J. Bacteriol,* 174, 4026–4035.

Davies, D. G., Parsek, M. R., Pearson, J. P., Iglewski, B. H., Costerton, J. W. and Greenberg, E. P. (1997). The involvement of cell-to-cell signals in the development of a bacterial biofilm. *Science,* 280, 295–298.

de Kievit, T. R., Seed, P. C., Passador, L., Nezezon, J. and Iglewski, B. H. (1999). RsaL, a novel repressor of virulence gene expression in *Pseudomonas aeruginosa*. *J. Bacteriol,* 181, 2175–2184.

de Kievit, T.R. and Iglewski, B.H. (2000). Bacterial quorum sensing in pathogenic relationships. *Infect Immun,* 68, 4839–4849.

Diggle, S.P., Winzer, K., Lazdunski, A., Williams, P. and Camara, M. (2002). Advancing the quorum in *Pseudomonas aeruginosa*: MvaT and the regulation of N-acyl homoserine lactone production and virulence gene expression. *J Bacteriol,* 184, 2576–2586.

Duan, K., Dammel, C., Stein, J., Rabin, H., and Surette, M.G. (2003). Modulation of *Pseudomonas aeruginosa* gene expression by host microflora through interspecies communication. *Mol Microbiol,* 50, 1477–1491.

Dunn, M., and Wunderink, R. G. (1995). Ventilator-associated pneumonia caused by *Pseudomonas* infection. *Clinics Chest Med,* 16, 95–109.

Eberl, L. and Tümmler, B. (2004). *Pseudomonas aeruginosa* and *Burkholderia cepacia* in cystic fibrosis: genome evolution, interactions and adaptation. *Int J Med Microbiol,* 294, 123–131.

Eberl, L., Winson, M. K., Sternberg, C., Stewart, G. S. A. B., Christiansen, G., Chhabra, S. R., Bycroft, B. W., Williams, P., Molin, S. and Givskov, M. (1996). Involvement of *N*-acyl-L-homoserine lactone autoinducers in controlling the multicellular behavior of *Serratia liquifaciens*. *Mol. Microbiol,* 20, 127–136.

Engebrecht, J. and Silverman. M. (1987). Nucleotide sequence of the regulatory locus controlling expression of bacterial genes for bioluminescence. *Nucleic Acids Res.,* 15, 10455–10467.

Fagerlind, M.G., Rice, S.A., Nilsson, P., Harlen, M., James, S., Charlton, T. and Kjelleberg, S. (2004). The role of regulators in the expression of quorum- sensing signals in *Pseudomonas aeruginosa. J. Mol. Microbiol Biotechnol,* 6, 88–100.

Fergie, J. E., Shema, S. J., Lott, L., Crawford, R. and Patrick, C. C. (1994). *Pseudomonas aeruginosa* bacteremia in immunocompromised children: analysis of factors associated with a poor outcome. *Clin. Infect. Dis.,* 18, 390–394.

Flavier, A. B., Ganova-Raeva, L. M., Schell, M. A. and Denny, T. P. (1997). Hierarchical autoinduction in *Ralstonia solanacearum*: control of acyl-homoserine lactone production by a novel autoregulatory system responsive to 3-hydroxypalmitic acid ester. *J. Bacteriol,* 179, 7089–7097.

Fuqua, C. and Greenberg, E.P. (2002). Listening on bacteria: acyl-homoserine lactone signalling. *Nat Rev Mol Cell Biol,* 3, 685–695.

Gambello, M. J. and Iglewski, B. H. (1991). Cloning and characterization of the *Pseudomonas aeruginosa lasR* gene: a transcriptional activator of elastase expression. *J. Bacteriol,* 173, 3000–3009.

Gambello, M. J., Kaye, S. and Iglewski, B. H. (1993). LasR of *Pseudomonas aeruginosa* is a transcriptional activator of the alkaline protease gene (*apr*) and an enhancer of exotoxin A expression. *Infect. Immun.,* 61, 1180–1184.

Givskov, M., Eberl, L. and Molin, S. (1997). Control of exoenzyme production, motility and cell differentiation in *Serratia liquifaciens. FEMS Microbiol. Lett,* 148, 115–122.

Gray, K. M., Pearson, J. P., Downie, J. A., Boboye, B. E. A. and Greenberg, E. P. (1996). Cell-to-cell signaling in the symbiotic nitrogen-fixing bacterium *Rhizobium leguminosarum*: autoinduction of stationary phase and rhizosphere-expressed genes. *J. Bacteriol,* 178, 372–376.

Greenberg, E.P. (2003). Tiny teamwork. *Nature,* 424, 134.

Hassett, D.J., Ma, J.F., Elkins, J.G., McDermott, T.R., Ochsner, U.A., West, S.E.H., et al. (1999). Quorum sensing in *Pseudomonas aeruginosa* controls expression of catalase and superoxide dismutase genes and mediates biofilm susceptibility to hydrogen peroxide. *Mol Microbiol,* 34, 1082–1093.

Hentzer, M. and Givskov, M. (2003). Pharmacological inhibition of quorum sensing for the treatment of chronic bacterial infections. *J. Clin. Invest,* 112, 1300–1307.

Hentzer, M., Givskov, M. and Eberl, L. (2004). Quorum sensing in Biofilms: Gossip in slime city. In *Microbial Biofilms*. O'Toole, G. and Ghannoum, M.A. (eds). Washington, DC, USA: American Society for Microbiology Press, pp. 118–140.

Heurlier, K., Denervaud, V., Pessi, G., Reimmann, C. and Haas, D. (2003). Negative control of quorum sensing by RpoN (sigma 54) in *Pseudomonas aeruginosa* PAO1. *J. Bacteriol.,* 185, 2227–2235.

Heurlier, K., Williams, F., Heeb, S., Dormond, C., Pessi, G., Singer, D., et al. (2004). Positive control of swarming, rhamnolipid synthesis and lipase production by the posttranscriptional RsmA/RsmZ system in *Pseudomonas aeruginosa* PAO1. *J. Bacteriol.,* 186, 2936–2945.

Hogardt, M., Roeder, M., Schreff, A.M., Eberl, L., and Heesemann, J. (2004) Expression of *Pseudomonas aeruginosa* exoS is controlled by quorum sensing and RpoS. *Microbiology* 150: 843–851.

Hwang, I., Pei-Li, L., Zhang, L., Piper, K. R., Cook, D. M., Tate, M. E. and Farrand, S. K. (1994). TraI, a LuxI homolog, is responsible for production of conjugation factor, the Ti plasmid *N*-acyl homoserine lactone autoinducer. *Proc. Natl. Acad. Sci. USA,* 91, 4639–4643.

Jones, S., Yu, B., Bainton, N. J., Birdsall, M., Bycroft, B. W., Chhabra, S. R., Cox, A. J. R., Golby, P., Reeves, P. J., Stephens, S., Winson, M. K., Salmond, G. P. C., Stewart, G. S. A. B. and Williams, P. (1993). The *lux* autoinducer regulates the production of exoenzyme virulence determinants in *Erwinia carotovora* and *Pseudomonas aeruginosa. EMBO J.,* 12, 2477–2482.

Jude, F., Kohler, T., Branny, P., Perron, K., Mayer, M.P., Comte, R. and van Delden, C. (2003). Posttranscriptional control of quorum-sensing-dependent virulence genes by DksA in *Pseudomonas aeruginosa. J. Bacteriol.,* 185, 3558–3566.

Juhas, M., Wiehlmann, L., Huber, B., Jordan, D., Lauber, J., Salunkhe, P., et al. (2004). Global regulation of quorum sensing and virulence by VqsR in *Pseudomonas aeruginosa. Microbiology,* 150, 831–841.

Kempner, E.S. and Henson, F.E. (1968). Aspects of light production from *photobacterium fischeri. J. Bacteriol.*, 109, 1101-1105.

Kohler, T., Michea-Hamzehpour, M., Henze, U., Gotoh, N., Curty, L. K. and Pechere, J. C. (1997). Characterization of MexE-MexF-OprN, a positively regulated multidrug efflux system of *Pseudomonas aeruginosa. Mol. Microbiol.,* 23, 345–354.

Latifi, A., Foglino, M., Tanaka, K., Williams, P. and Lazdunski, A. (1996). A hierarchical quorum-sensing cascade in *Pseudomonas aeruginosa* links the transcriptional activators LasR and RhlR (VsmR) to expression of the stationary phase sigma factor RpoS. *Mol. Microbiol.,* 21, 1137–1146.

Latifi, A., Winson, M. K., Foglino, M., Bycroft, B. W., Stewart, G. S. A. B., Lazdunski, L. and Williams, P. (1995). Multiple homologues of LuxR and LuxI control expression of virulence determinants and secondary metabolites through quorum sensing in *Pseudomonas aeruginosa* PAO1. *Mol. Microbiol.* 17, 333–343.

Lazdunski, A.M., Ventre, I. and Sturgis, J.N. (2004). Regulatory circuits and communication in gram-negative bacteria. *Nat. Rev. Microbiol.,* 7, 581–592.

Ledgham, F., Soscia, C., Chakrabarty, A., Lazdunski, A. and Foglino, M. (2003a). Global regulation in *Pseudomonas aeruginosa:* the regulatory protein AlgR2 (AlgQ) acts as a modulator of quorum sensing. *Res. Microbiol.,* 154, 207–213.

Ledgham, F., Ventre, I., Soscia, C., Foglino, M., Sturgis, J.N. and Lazdunski, A. (2003b). Interaction of the quorum sensing regulator QscR: interaction with itself and the other regulators of *Pseudomonas aeruginosa* LasR and RhlR. *Mol. Microbiol.,* 48, 199–210.

Lewenza, S., Conway, B., Greenberg, E. P. and Sokol, P. A. (1999). Quorum sensing in *Burkholderia cepacia*: identification of the LuxRI homologs CepRI. *J. Bacteriol.,* 181, 748–756.

Luo, P. and Morrison, D. A. (2003). Transient association of an alternative sigma factor, ComX, with RNA polymerase during the period of competence for genetic transformation in *Streptococcus pneumoniae. J. Bacteriol,* 185, 349–58.

Luo, P., Li, H. and Morrison, D.A. (2003). ComX is a unique link between multiple quorum sensing outputs and competence in *Streptococcus pneumoniae*. *Mol. Microbiol.,* 50, 623–33.

McGowan, S., Sebaihia, M., Jones, S., Yu, S., Bainton, N., Chan, P. F., Bycroft, B. W., Stewart, G. S. A. B., Salmond, G. P. C. and Williams, P. (1995). Carbapenem antibiotic production in *Erwinia carotovora* is regulated byCarR, a homologue of the LuxR transcriptional activator. *Microbiology*, 141, 541–550.

McLean, K. H., Winson, M. K., Fish, L., Taylor, A., Chhabra, S. R., Camara, M., Daykin, M., Lamb, J. H., Swift, S., Bycroft, B. W., Stewart, G. S. A. B. and Williams, P. (1997). Quorum sensing and *Chromobacterium violaceum*: exploitation of violacein production and inhibition for the detection of *N*-acyl homoserine lactones. *Microbiology*, 143, 3703–3711.

Mendelson, M. H., Gurtman, A., Szabo, S., Neibart, E., Meyers, B. R., Policar, M., Cheung, T. W., Lillienfeld, D., Hammer, G., Reddy, S., Choi, K. and Hirschman, S. Z. (1994). *Pseudomonas aeruginosa* bacteremia in patients with AIDS. *Clin. Infect. Dis*, 18, 886–895.

Miller, M.B. and Bassler, B.L. (2001). Quorum sensing in bacteria. *Annu. Rev. Microbiol.,* 55, 165–169.

Milton, D. L., Hardman, A., Camara, M., Chhabra, S. R., Bycroft, B. W., Stewart, G. S. A. B. and Williams, P. (1997). Quorum sensing in *Vibrio anguillarum*: characterization of the *vanI/vanR* locus and identification of the autoinducer *N*-(3-oxododecanoyl)-L-homoserine lactone. *J. Bacteriol.,* 179, 3004–3012.

Nasser, W., Bouillant, M. L., Salmond, G. and Reverchon, S. (1998). Characterization of the *Erwinia chrysanthemi expI-expR* locus directing the synthesis of two *N*-acyl-homoserine lactone signal molecules. *Mol. Microbiol*, 29, 1391–1405.

Nealson, K. H. (1977). Autoinduction of bacterial luciferase: Occurrence, mechanism and significance. *Arch. Microbiol,* 112, 73–79.

Ochsner, U. A. and Reiser, J. (1995). Autoinducer-mediated regulation of rhamnolipid biosurfactant synthesis in *Pseudomonas aeruginosa*. *Proc. Natl. Acad. Sci. USA,* 92, 6424–6428.

Ochsner, U. A., Koch, A. K., Fiechter, A. and Reiser, J. (1994). Isolation and characterization of a regulatory gene affecting rhamnolipid biosurfactant synthesis in *Pseudomonas aeruginosa*. *J. Bacteriol.* 176, 2044–2054.

Pappas, K.M., Weingart, C.L., and Winans, S.C. (2004). Chemical communication in proteobacteria: biochemical and structural studies on signal synthases and receptors required for intracellular signalling. *Mol. Microbiol.,* 53, 755–769.

Passador, L., Cook, J. M., Gambello, M. J., Rust, L. and Iglewski, B. H. (1993). Expression of *Pseudomonas aeruginosa* virulence genes requires cell-to-cell communication. *Science*, 260, 1127–1130.

Pearson, J. P., Gray, K. M., Passador, L., Tucker, K. D., Eberhard, A., Iglewski, B. H. and Greenberg, E. P. (1994). Structure of the autoinducer required for expression of *Pseudomonas aeruginosa* virulence genes. *Proc. Natl.Acad. Sci. USA,* 91, 197–201.

Pearson, J. P., Passador, L., Iglewski, B. H. and Greenberg, E. P. (1995). A second *N*-acyl homoserine lactone signal produced by *Pseudomonas aeruginosa*. *Proc. Natl. Acad. Sci. USA*, 92, 1490–1494.

Pearson, J. P., Pesci, E. C. and Iglewski, B. H. (1997). Roles of *Pseudomonas aeruginosa las* and *rhl* quorum-sensing systems in control of elastase and rhamnolipid biosynthesis genes. *J. Bacteriol.,* 179, 5756–5767.

Pearson, J. P., Van Delden, C. and Iglewski, B. H. (1999). Active efflux and diffusion are involved in transport of *Pseudomonas aeruginosa* cell-to-cell signals. *J. Bacteriol.,* 181, 1203–1210.

Pesci, E. C., Milbank, J. B., Pearson, J. P., McKnight, S., Kende, A. S., Greenberg, E. P. and Iglewski, B. H. (1999). Quinolone signaling in the cell-to-cell communication system of *Pseudomonas aeruginosa*. *Proc. Natl. Acad. Sci. USA*, 96, 11229–11234.

Pesci, E. C., Pearson, J. P., Seed, P. C. and Iglewski, B. H. (1997). Regulation of *las* and *rhl* quorum sensing in *Pseudomonas aeruginosa*. *J. Bacteriol.,* 179, 3127–3132.

Pessi, G. and Haas, D. (2000). Transcriptional control of the hydrogen cyanide biosynthetic genes *hcnABC* by the anaerobic regulator ANR and the quorum-sensing regulators LasR and RhlR in *Pseudomonas aeruginosa*. *J. Bacteriol.,* 182, 6940–949.

Pessi, G., Williams, F., Hindle, Z., Heurlier, K., Holden, M.T.G., Camara, M., et al. (2001). The global posttranscriptional regulator RsmA modulates production of virulence determinants and *N*-acylhomoserine lactones in *Pseudomonas aeruginosa*. *J. Bacteriol.,* 183, 6676–6683.

Piper, K. R., Beck von Bodman, S. and Farrand, S. K. (1993). Conjugation factor of *Agrobacterium tumefaciens* regulates Ti plasmid transfer by autoinduction. *Nature,* 362, 448–450.

Pirhonen, M., Flego, D., Heikinheimo, R. and Palva, E. T. (1993). A small diffusible signal molecule is responsible for the global control of virulence and exoenzyme production in the plant pathogen *Erwinia carotovora*. *EMBO J.*, 12, 2467–2476.

Poole, K., Gotoh, N., Tsujimoto, H., Zhao, Q., Wada, A., Yamasaki, T., Neshat, S., Yamagishi, J., Li, X. Z. and Nishino, T. (1996). Overexpression of the *mexC-mexC-oprJ* efflux operon in *nfxB*-type multidrug-resistant strains of *Pseudomonas aeruginosa*. *Mol. Microbiol.,* 21, 713–724.

Poole, K., Krebes, K., McNally, C. and Neshat, S. (1993). Multiple antibiotic resistance in *Pseudomonas aeruginosa*: evidence for involvement of an efflux operon. *J. Bacteriol.,* 175, 7363–7372.

Puskas, A., Greenberg, E. P., Kaplan, S. and Schaefer, A. L. (1997). A quorum-sensing system in the free-living photosynthetic bacterium *Rhodobacter sphaerhoides*. *J. Bacteriol.,* 179, 7530–7537.

Rahme, L. G., Stevens, E. J., Wolfort, S. F., Shao, J., Tompkins, R. G. and Ausubel, F. M. (1995). Common virulence factors for bacterial pathogenicity in plants and animals. *Science,* 268, 1899–1902.

Rahme, L. G., Tan, M.-W., Le, L., Wong, S. M., Tompkins, R. G., Calderwood, S. B. and Ausubel, F. M. (1997). Use of model plant hosts to identify *Pseudomonas aeruginosa* virulence factors. *Proc. Natl. Acad. Sci. USA*, 94, 13245–13250.

Rahme, L.G., Stevens, E.J., Wolfort, S.F., Shao, J., Tompkins, R.G. and Ausubel, F.M. (1995). Common virulence factors for bacterial pathogenicity in plants and animals. *Science,* 268, 1899–1902.

Rashid, M.H., Rumbaugh, K., Passador, L., Davies, D.G., Hamood, A.N., Iglewski, B.H. and Kornberg, A. (2000). Polyphosphate kinase is essential for biofilm development, quorum sensing and virulence of *Pseudomonas aeruginosa. Proc. Natl. Acad. Sci. USA,* 97, 9636–9641.

Reimmann, C., Beyeler, M., Latifi, A., Winteler, H., Foglino, M., Lazdunski, A. and Haas, D. (1997). The global activator GacA of *Pseudomonas aeruginosa* PAO positively controls the production of the autoinducer *N*-butyrylhomoserine lactone and the formation of the virulence factors pyocyanin, cyanide, and lipase. *Mol. Microbiol,* 24, 309–319.

Reverchon, S., Bouillant, M. L., Salmond, G. and Nasser, W. (1998). Integration of the quorum-sensing system in the regulatory networks controlling virulence factor synthesis in *Erwinia chrysanthemi. Mol. Microbiol.,* 29, 1407–1418.

Rodelas, B., Lithgow, J. K., Wisniewski-Dye, F., Hardman, A., Wilkinson, A., Economou, A., Williams, P. and Downie, J. A. (1999). Analysis of quorum sensing-dependent control of rhizosphere-expressed (*rhi*) genes in *Rhizobium leguminosarum* bv. *viciae. J. Bacteriol.*, 181, 3816–3823.

Rosemeyeer, V., Michiels, J., Verreth, C. and Vanderleyden, J. (1998). *luxI* and *luxR*-homologous genes of *Rhizobium etli* CNPAF512 contribute to synthesis of autoinducer molecules and nodulation of *Phaseolus vulgaris. J. Bacteriol,* 180, 815–821.

Rumbaugh, K.P., Griswold, J.A. and Hamood, A.N. (1999). Contribution of the regulatory gene lasR to pathogenesis of *Pseudomonas aeruginosa* infection of burned mice. *J. Burn Care Rehabil.,* 20, 42–49.

Schuster, M., Hawkins, A.C., Harwood, C.S. and Greenberg, E.P. (2004). The *Pseudomonas aeruginosa* RpoS regulon and its relationship to quorum sensing. *Mol. Microbiol.,* 51, 973–985.

Schuster, M., Lostroh, C.P., Ogi, T. and Greenberg, E.P. (2003). Identification, timing and signal specificity of *Pseudomonas aeruginosa* quorum-controlled genes: a transcriptome analysis. *J. Bacteriol.,* 185, 2066–2079.

Seed, P. C., Passador, L. and Iglewski, B. H. (1995). Activation of the *Pseudomonas aeruginosa lasI* gene by LasR and the *Pseudomonas* autoinducer PAI: an autoinduction regulatory hierarchy. *J. Bacteriol.,* 177, 654–659.

Sharma, S., Stark, T. F., Beattie, W. G. and Moses, R. E. (1986). Multiple control elements for the *uvrC* gene unit of *Escherichia coli. Nucleic Acids Res,* 14, 2301–2318.

Showalter, R. E., Martin, M. O. and Silverman, M. R. (1990). Cloning and nucleotide sequence of *luxR*, a regulatory gene controlling luminescence in *Vibrio harveyi. J. Bacteriol.,* 172, 2946–2954.

Smith, R.S., Wolfgang, M.C. and Lory, S. (2004). An adenylate cyclase-controlled signalling network regulates *Pseudomonas aeruginosa* virulence in mouse model of acute pneumonia. *Infect Immun.,* 72, 1677–1684.

Sperandio, V., Mellies, J. L., Nguyen, W., Shin, S. and Kaper, J. B. (1999). Quorum sensing controls expression of the type III secretion gene transcription and protein secretion in enterohemorrhagic and enteropathogenic *Escherichia coli. Proc. Natl. Acad. Sci. USA*, 96, 15196–15201.

Swift, S., Karlyshev, A. V., Durant, E. L., Winson, M. K., Chhabra, S. R., Williams, P., Macintyre, S. and Stewart, G. S. A. B. (1997). Quorum sensing in *Aeromonas hydrophila* and *Aeromonas salmonicida*: identification of the LuxRI homologues AhyRI and AsaRI and their cognate signal molecules. *J. Bacteriol.,* 179, 5271–5281.

Swift, S., Winson, M. K., Chan, P. F., Bainton, N. J., Birdsall, M., Reeves, P. J., Rees, C. E. D., Chhabra, S. R., Hill, P. J., Throup, J. P., Bycroft, B. W., Salmond, G. P. C., Williams,

P. and Stewart, G. S. A. B. (1993). A novel strategy for the isolation of *luxI* homologues: evidence for the widespread distribution of a LuxR:LuxI superfamily in enteric bacteria. *Mol. Microbiol,* 10, 511–520.

Tan, M.W., Rahme, L.G., Sternberg, J.A., Tompkins, R.G. and Ausubel, F.M. (1999). *Pseudomonas aeruginosa* killing of *Caenorhabditis elegans* used to identify *P. aeruginosa* virulence factors. *Proc. Natl. Acad. Sci. USA,* 96, 2408–2413.

Tang, H.B., DiMango, E., Bryan, R., Gambello, M., Iglewski, B.H., Goldberg, J.B. and Prince, A. (1996). Contribution of specific *Pseudomonas aeruginosa* virulence factors to pathogenesis of pneumonia in a neonatal mouse model of infection. *Infect Immun.,* 64, 37–43.

Thompson, L.S., Webb, J.S., Rice, S.A. and Kjelleberg, S. (2003). The alternative sigma factor RpoN regulates the quorum sensing gene rhlI in *Pseudomonas aeruginosa. FEMS Microbiol. Lett,* 220, 187–195.

Throup, J. P., Camara, M., Briggs, G. S., Winson, M. K., Chhabra, S. R., Bycroft, B. W., Williams, P. and Stewart, G. S. A. B. (1995). Characterization of the *yenI/R* locus from *Yersinia enterocolitica* mediating the synthesis of two *N*-acyl-homoserine lactone signal molecules. *Mol. Microbiol.,* 17, 345–356.

Toder, D. S., Gambello, M. J. and Iglewski, B. H. (1991). *Pseudomonas aeruginosa* LasA: a second elastase gene under transcriptional control of *lasR. Mol. Microbiol.,* 5, 2003–2010.

Tümmler, B., Koopman, U., Grothues, D., Weissbrodt, H., Steinkamp, D. and von der Hardt, H. (1991). Nosocomial acquisition of *Pseudomonas aeruginosa* by cystic fibrosis patients. *J. Clin. Microbiol.,* 29,1265–1267.

Vasil, M.L. (2003). NA Microarrays in analysis of *quorum Sensing* strengths and limitations. *J. Bacteriol.,* 185, 2061-2065.

Wagner, V.E., Bushnell, D., Passador, L., Brooks, A.I. and Iglewsi, B.H. (2003). Microarray analysis of *Pseudomonas aeruginosa* quorum-sensing regulons: effects of growth phase and environment. *J. Bacteriol.,* 185, 2080–2095.

Whitehead, N.A., Barnard, M.L., Slater, H., Simpson, N.J.L. and Salmond, G.P.C. (2001). Quorum-sensing in Gram negative bacteria. *FEMS Microbiol Rev,* 25, 365–404.

Whiteley, M., Parsek, M.R. and Greenberg, E.P. (2000). Regulation of quorum sensing by RpoS in *Pseudomonas aeruginosa. J. Bacteriol.,* 182, 4356–4360.

Winson, M. K., Camara, M., Latifi, A., Foglino, M., Chhabra, S. R., Daykin, M., Bally, M., Chapon, V., Salmond, G. P. C., Bycroft, B. W., Lazdunski, A., Stewart, G. S. A. B. and Williams, P. (1995). Multiple *N*-acyl-*L*-homoserine lactone signal molecules regulate production of virulence determinants and secondary metabolites in *Pseudomonas aeruginosa. Proc. Natl. Acad. Sci. USA,* 92, 9427–9431.

Winzer, K., Falconer, C., Garber, N.C., Diggle, S.P., Camara, M. and Williams, P. (2000). The *Pseudomonas aeruginosa* lectins PA-IL and PA-IIL are controlled by quorum sensing and by RpoS. *J. Bacteriol.,* 182, 6401–6411.

Withers, H. L. and Nordstrom, K. (1998). Quorum-sensing acts at initiation of chromosomal replication in *Escherichia coli. Proc. Natl. Acad. Sci. USA,* 95, 15694–15699.

Wolfgang, M.C., Lee, V.T., Gimore, M.E. and Lory, S. (2003). Coordinate regulation of bacterial virulence genes by a novel adenylate cyclase-dependent signalling pathway. *Dev. Cell,* 4, 253–263.

Wu, Y.P., McMahon, E., Kraine, M.R., Tisch, R., Meyers, A., Frelinger, J., et al. (2000). Distribution and characterization of GFP(+) donor hematogenous cells in Twitcher mice after bone marrow transplantation. *Am. J. Pathol.,* 156, 1849–1854.

Zhang, L., Murphy, P. J., Kerr, A. and Tate, M. E. (1993). *Agrobacterium* conjugation and gene regulation by *N*-acyl-L-homoserine lactones. *Nature,* 362, 446–448.

Index

D

I

J

K

L

M

N

O

P

Q

R

S

T

U

V